Author's Note

Textbook authors would present the subject of study from their scholarly perspective with an emphasis on pedagogy. As an alternative, this book took care to preserve the student's view on the subject by faithfully reproducing class notes along with improvements and extensions made to the notes even long after the student had passed the examination. The approaches of presenting the subject complement each other and help to achieve the objective of relaying the information to the viewer. As an effort to entertain the viewer, the main text is followed by one or two selected pdf-files, collected earlier from the now expired (defunct) website www.2btvd.net that over the years of its existence hosted much of the material presented.

Melbourne, in March 2024

Synopsis

Was it a god (Boltzmann's saying) that wrote those equations? Such was the enthusiasm for the Maxwell Equations. A review of Electromagnetic Field Theory (EMF), the repetitorium refers to the Hertz vector as the basis for discussions. The treatment of EMF assumes knowledge of coordinate systems along with scalar and vector fields. Electric and magnetic field strengths are computed from the Hertz vector for the coordinate system chosen. As an alternative the approach using Green's Theorem is discussed. A favorite topics of electrical engineering, antennas tend to attract great interest from students, which encourages one to look back at the elementary dipole as the basic element for antenna designs. From field strengths and their patterns to eigenfrequencies, the topics flow naturally from one section to the next. The less so common 4-dimensional spacetime formulation is also touched on as an addendum. One hundred fifty years on since Maxwell, the repetitorium is echoing the enthusiasm that was ringing among scientists.

Nomenclature and Symbols

Samples of symbols listed in the order of functional groupings
(likely found in some functional relations with symbols shown in their vicinity)

		MKSA system units	**Comment/Example**
Maxwell Equations	$\mathbf{H}$	Magnetic field strength A/m	TM transversal magnetic: Zero $\mathbf{H}$ component in the direction of propagation
	$\mathbf{E}$	Electric field strength V/m	TE: transversal electric
	$\mathbf{J}$	Conductive current density A/m^2	
	$\mathbf{B}$	Magnetic induction Vs/m^2	$\Phi = \int_A \mathbf{B}\,d\mathbf{A}$ Φ magnetic flow (number of magnetic field lines through A)
	ρ	Charge density As/m^3	
	μ	Permeability Vs/Am	
	ε	Permittivity As/Vm (dielectric constant)	
Basic math	$(grad\varphi)_n$	Gradient of scalar φ projected in direction n	$(\nabla\varphi)_n = \lim_{dn\to 0}\frac{\varphi(\mathbf{x}+d\mathbf{x})-\varphi(\mathbf{x})}{dn}$
	div	Divergence operator	
	$curl$	Curl operator	
	$\delta(.)$	Dirac impulse	
	$\frac{d}{dt}$, $\frac{\partial}{\partial t}$	Derivative, Partial derivative	$\frac{\partial}{\partial t}$ sometimes written as $\frac{\delta}{\delta t}$
	$\sqrt{\sum_{i=1}^{3} ds_i^2}$	Elementary geometric distance	$ds_i = g_i\,dx_i$, $i = 1,2,3$ Metric coefficients g_i
	∇	Nabla operator (Hamilton operator)	$grad\varphi = \nabla\varphi$, $div\,\mathbf{v} = \nabla\mathbf{v}$, $curl\,\mathbf{v} = \nabla\times\mathbf{v}$
	$div\ grad \equiv \Delta$	Laplace operator	The differential equation for the spherical Hertz vector component Π_r, and that for a scalar ψ (i.e. wave equation $\Delta\psi + k^2\psi = 0$ in spherical coordinates) may look similar, but they are not the same
	$(.,.)$	Inner product on inner product spaces i.e. orthogonal spaces (scalar product), refer to operator theory for Hilbert spaces etc.	$(L^*v,u)_1 = (v,Lu)_2$ definition of adjoint L^*, inner products on spaces 1 and 2 respectively If $L^*L = LL^*$, operator called *normal* If $L^* = L$, operator called *self-adjoint*

Formulation for the Hertz vector	$\mathbf{A}$	Vector potential	
Formulation for the Hertz vector	$\mathbf{\Pi}_e$, $\mathbf{\Pi}_m$	Electric, magnetic Hertz vector	
Formulation for the Hertz vector	$J_n(kr)$, $N_n(kr)$	Bessel, Neumann function of order n — Generic symbol $Z_n(kr)$	Hankel functions $H_n^{(1)(2)}(kr) = J_n(kr) \pm jN_n(kr)$
Formulation for the Hertz vector	$P_n^m(x)$	Associated Legendre function of order n of index m	Symbol $P_n^m(x) \equiv (1-x)^{m/2} P_n^{(m)}(x)$ $P_n^{(m)}(x) \equiv \frac{d^m}{dx^m} P_n(x)$ m-th order derivative
Formulation for the Hertz vector	$P_n(x)$	Solutions to Legendre differential equation representable as infinite series	Depending on the order n, series reduce to polynomials known as Legendre polynomials (index m=0)
Maxwell Equations in the language of Mechanics	$\mathcal{L}$	Lagrange density function (Base function)	
Maxwell Equations in the language of Mechanics	$\mathcal{H}$	Hamilton density function	
Maxwell Equations in the language of Mechanics	δ	Variation symbol [$\frac{d}{dx}\delta y = \delta \frac{d}{dx} y = \delta y'$ δ commutes in this sense, for example $\delta(grad\ \varphi) = grad\ \delta\varphi$]	$J = \int_{t_1}^{t_2} F dt$; $\delta J = \delta \int_{t_1}^{t_2} F dt = 0$ Variation of the integral of F to be 0 Assuming $F(t, q, \dot{q})$, the Euler differential equation would read $\frac{\partial F}{\partial q} - \frac{d}{dt}\frac{\partial F}{\partial \frac{\delta q}{\delta t}} = 0$, the solution to which would satisfy $\delta J = 0$ regardless of δq, achieving an extremum max. or min. for J
Maxwell Equations in the language of Mechanics	η, $\mathcal{P}$	Generalized coordinates, Impulse coordinates	$\mathcal{P} = \frac{\partial \mathcal{L}}{\partial \dot{\eta}}$, $\dot{\mathcal{P}} = -\frac{\delta \mathcal{H}}{\delta \eta}$, $\dot{\eta} = \frac{\delta \mathcal{H}}{\delta \mathcal{P}}$

§§§

Table of Contents

EMF Equations: An Overview

General perspective

Network theory	⇔ (Analogy)	Point mechanics
⇓		⇓
		1st quantum step
Electromagnetic field theory		Quantum mechanics
⇓		
4-dim spacetime formalism		
⇓		⇓
1st quantum step Photon field		*2nd quantum step* Electron-positron field
	⇓	
	Quantum electrodynamics	

A view on engineering and science, the table depicts progresses made over the last century. The diverse branches going from network theory to quantum mechanics are characterized as particular disciplines of science. Relativistic electrodynamics achieved *Vollkommenheit* (perfection) and *Abgeschlossenheit* (completion), to quote from a textbook, completion or completeness meaning in everyone's language that not a single little stone had been left unturned. This repetitorium restricts itself on electromagnetic field theory as part of the bigger picture.

Maxwell field equations

Homogeneous media characterized by the constants ε and μ [Simonyi, Küpfmüller]

I	$curl\,\mathbf{H} = \mathbf{J} + \varepsilon\dfrac{\partial\mathbf{E}}{\partial t}$	III	$div\,\mathbf{H} = 0$
II	$curl\,\mathbf{E} = \quad -\mu\dfrac{\partial\mathbf{H}}{\partial t}$	IV	$div\,\mathbf{E} = \dfrac{\rho}{\varepsilon}$

$\dfrac{\partial}{\partial t}$	**J**	**Case**				**Comment**
0	0	I	$curl\,\mathbf{H} = 0$	III	$div\,\mathbf{H} = 0$	Electro statics
		II	$curl\,\mathbf{E} = 0$	IV	$div\,\mathbf{E} = \dfrac{\rho}{\varepsilon}$	
0	Non-zero	I	$curl\,\mathbf{H} = \mathbf{J}$	III	$div\,\mathbf{H} = 0$	Stationary
		II	$curl\,\mathbf{E} = 0$	IV	$div\,\mathbf{E} = \dfrac{\rho}{\varepsilon}$	
Non-zero	0	I	$curl\,\mathbf{H} = \varepsilon\dfrac{\partial\mathbf{E}}{\partial t}$	III	$div\,\mathbf{H} = 0$	Non-zero displacement current density
		II	$curl\,\mathbf{E} = \quad -\mu\dfrac{\partial\mathbf{H}}{\partial t}$	IV	$div\,\mathbf{E} = \dfrac{\rho}{\varepsilon}$	
Non-zero	Non-zero	I	$curl\,\mathbf{H} = \mathbf{J}$	III	$div\,\mathbf{H} = 0$	Quasi stationary
		II	$curl\,\mathbf{E} = \quad -\mu\dfrac{\partial\mathbf{H}}{\partial t}$	IV	$div\,\mathbf{E} = \dfrac{\rho}{\varepsilon}$	
		I	$curl\,\mathbf{H} = \mathbf{J} + \varepsilon\dfrac{\partial\mathbf{E}}{\partial t}$	III	$div\,\mathbf{H} = 0$	General case
		II	$curl\,\mathbf{E} = \quad -\mu\dfrac{\partial\mathbf{H}}{\partial t}$	IV	$div\,\mathbf{E} = \dfrac{\rho}{\varepsilon}$	

It would be beyond the scope of this repetitorium to extend to 4-dimensional generalized spacetime variables for the treatment of electromagnetic field theory, the subject of this text. Rather, evolved from student class notes, the repetitorium places an emphasis on the formulation for the Hertz vector $\mathbf{\Pi}_e$ (or $\mathbf{\Pi}_m$) as the basis for discussions. The geometry at hand specifies the choice for the coordinate system in which to express the results.

Separation of variables turned out to be of great importance to the classical treatment of EMF. The method led to Bessel and Legendre differential equations for, respectively, cylindrical and spherical coordinates. In the next section, it may be worth recollecting one or two things about general coordinates.

General Coordinates

Expression for the elementary distance

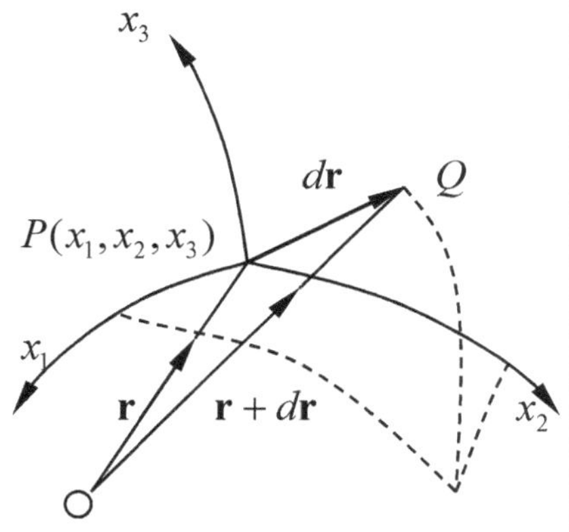	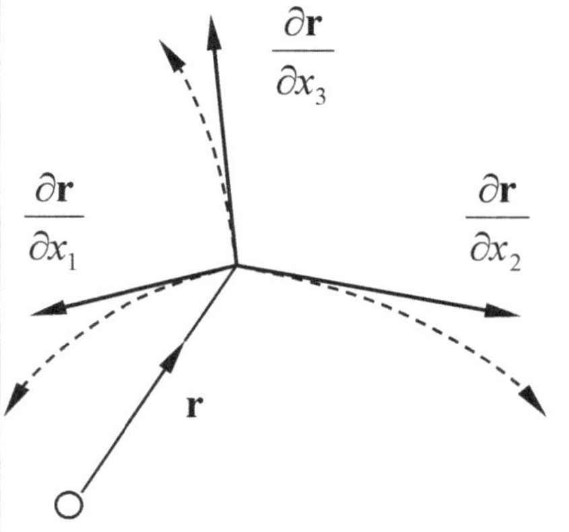	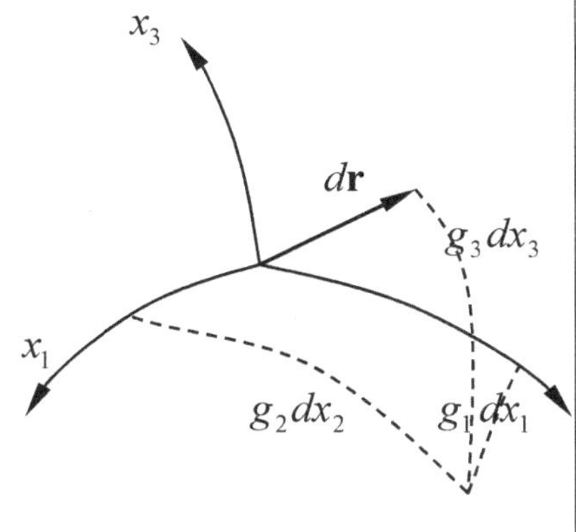
Elementary Distance between $P(x_1,x_2,x_3)$ and $Q(x_1+dx_1,x_2+dx_2,x_3+dx_3)$ is $d\mathbf{r}=(dx,dy,dz)$ There is a mapping from $d\mathbf{x}=(dx_1,dx_2,dx_3)$ to $d\mathbf{r}=(dx,dy,dz)$	Tangents of the coordinates' axes at an arbitrary point $P(x_1,x_2,x_3)$ Orthogonality is assumed for further proceedings	Elementary Geometric Distance is $\sqrt{\sum_{i=1}^{3} ds_i^2}$ where $ds_i = g_i\,dx_i$, $i=1,2,3$ See text for $g_1\,dx_1$, $g_2 dx_2$, $g_3 dx_3$

Distance between $P(x_1,x_2,x_3)$ and $Q(x_1+dx_1,x_2+dx_2,x_3+dx_3)$ is determined from $ds^2 = d\mathbf{r}.d\mathbf{r}$

$d\mathbf{r}=(dx,dy,dz)$ is given by $d\mathbf{x}=(dx_1,dx_2,dx_3)$ mapped by the tensor $\mathbf{T}_{dx}^{dr}$, namely $d\mathbf{r}=\mathbf{T}_{dx}^{dr}\,d\mathbf{x}$

$$\mathbf{T}_{dx}^{dr} = \begin{bmatrix} \frac{\partial x}{\partial x_1} & \frac{\partial x}{\partial x_2} & \frac{\partial x}{\partial x_3} \\ \frac{\partial y}{\partial x_1} & \frac{\partial y}{\partial x_2} & \frac{\partial y}{\partial x_3} \\ \frac{\partial z}{\partial x_1} & \frac{\partial z}{\partial x_2} & \frac{\partial z}{\partial x_3} \end{bmatrix}$$

$$d\mathbf{r} = \mathbf{T}_{dx}^{dr} d\mathbf{x} = \begin{bmatrix} \frac{\partial x}{\partial x_1} & \frac{\partial x}{\partial x_2} & \frac{\partial x}{\partial x_3} \\ \frac{\partial y}{\partial x_1} & \frac{\partial y}{\partial x_2} & \frac{\partial y}{\partial x_3} \\ \frac{\partial z}{\partial x_1} & \frac{\partial z}{\partial x_2} & \frac{\partial z}{\partial x_3} \end{bmatrix} \begin{bmatrix} dx_1 \\ dx_2 \\ dx_3 \end{bmatrix} = \begin{bmatrix} \frac{\partial x}{\partial x_1} dx_1 + \frac{\partial x}{\partial x_2} dx_2 + \frac{\partial x}{\partial x_3} dx_3 \\ \frac{\partial y}{\partial x_1} dx_1 + \frac{\partial y}{\partial x_2} dx_2 + \frac{\partial y}{\partial x_3} dx_3 \\ \frac{\partial z}{\partial x_1} dx_1 + \frac{\partial z}{\partial x_2} dx_2 + \frac{\partial z}{\partial x_3} dx_3 \end{bmatrix} \equiv \frac{\partial \mathbf{r}}{\partial x_1} dx_1 + \frac{\partial \mathbf{r}}{\partial x_2} dx_2 + \frac{\partial \mathbf{r}}{\partial x_3} dx_3$$

Proceeding with $ds^2 = d\mathbf{r}.d\mathbf{r}$

$$ds^2 = d\mathbf{r}.d\mathbf{r} = \left(\frac{\partial \mathbf{r}}{\partial x_1} dx_1 + \frac{\partial \mathbf{r}}{\partial x_2} dx_2 + \frac{\partial \mathbf{r}}{\partial x_3} dx_3 \right)^2$$

$$d\mathbf{r}.d\mathbf{r} = \left(\frac{\partial \mathbf{r}}{\partial x_1} \right)^2 dx_1^2 + \left(\frac{\partial \mathbf{r}}{\partial x_2} \right)^2 dx_2^2 + \left(\frac{\partial \mathbf{r}}{\partial x_3} \right)^2 dx_3^2 + 2\frac{\partial \mathbf{r}}{\partial x_1}\frac{\partial \mathbf{r}}{\partial x_2} dx_1 dx_2 + 2\frac{\partial \mathbf{r}}{\partial x_1}\frac{\partial \mathbf{r}}{\partial x_3} dx_1 dx_3 + 2\frac{\partial \mathbf{r}}{\partial x_2}\frac{\partial \mathbf{r}}{\partial x_3} dx_2 dx_3$$

The terms $\frac{\partial \mathbf{r}}{\partial x_1}$, $\frac{\partial \mathbf{r}}{\partial x_2}$ and $\frac{\partial \mathbf{r}}{\partial x_3}$ in the expression are recognized as the tangents of the coordinates (x_1, x_2, x_3) at the point $P(x_1, x_2, x_3)$ of interest. Only ***orthogonal*** coordinate systems are of interest. Consequently, the cross products between the terms $\frac{\partial \mathbf{r}}{\partial x_1}$, $\frac{\partial \mathbf{r}}{\partial x_2}$ and $\frac{\partial \mathbf{r}}{\partial x_3}$ yield zero.

$$\frac{\partial \mathbf{r}}{\partial x_1}\frac{\partial \mathbf{r}}{\partial x_2} = \frac{\partial \mathbf{r}}{\partial x_1}\frac{\partial \mathbf{r}}{\partial x_3} = \frac{\partial \mathbf{r}}{\partial x_2}\frac{\partial \mathbf{r}}{\partial x_3} \equiv 0$$

The expression becomes

$$d\mathbf{r}.d\mathbf{r} = \left(\frac{\partial \mathbf{r}}{\partial x_1} \right)^2 dx_1^2 + \left(\frac{\partial \mathbf{r}}{\partial x_2} \right)^2 dx_2^2 + \left(\frac{\partial \mathbf{r}}{\partial x_3} \right)^2 dx_3^2 \equiv g_1^2 dx_1^2 + g_2^2 dx_2^2 + g_3^2 dx_3^2$$

where g_i^2 has been written in place of the scalar product $(\partial \mathbf{r} / \partial x_i)^2$

$$g_i^2 \equiv \left(\frac{\partial \mathbf{r}}{\partial x_i} \right)^2 \qquad \text{for } i = 1,2,3$$

This is known as the Pythagore-an theorem applicable in small dimensions in the vicinity of the local point $P(x_1, x_2, x_3)$, involving the local distances $g_1 dx_1$, $g_2 dx_2$ and $g_3 dx_3$.

The unit vectors for the local Cartesian coordinate system at point $P(x_1, x_2, x_3)$ may be normalized to give

$$\frac{\partial \mathbf{r}}{\partial x_i} / \left| \frac{\partial \mathbf{r}}{\partial x_i} \right| \qquad \text{for } i = 1,2,3$$

whose (local) directions will change from point to point.

For any vector $\mathbf{v}(x_1, x_2, x_3)$ one may express its components as scalar products as follows

$$v_i(x_1, x_2, x_3) = \mathbf{v}(x_1, x_2, x_3) \cdot \frac{\partial \mathbf{r}}{\partial x_i} / \left| \frac{\partial \mathbf{r}}{\partial x_i} \right|$$

which represent the projections of the vector on the local coordinates.

The gradient in general coordinates

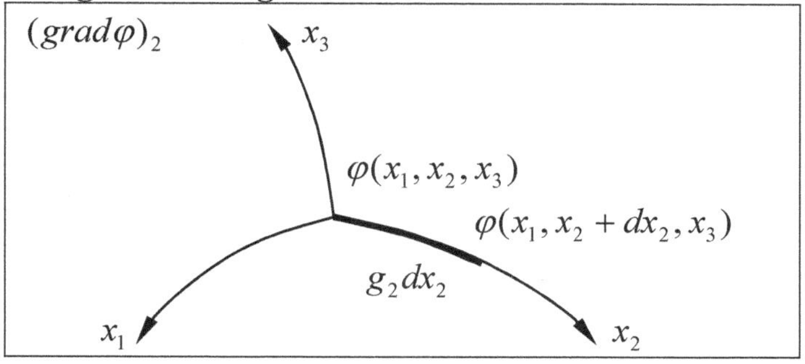

General definition for the gradient involves dividing by the infinitesimal *distance* dn

$$(grad\varphi)_n = \lim_{dn\to 0} \frac{\varphi(\mathbf{x} + d\mathbf{x}) - \varphi(\mathbf{x})}{dn}$$

applied to yield

$$(grad\varphi)_1 = \lim_{ds_1\to 0} \frac{\varphi(x_1 + dx_1, x_2, x_3) - \varphi(x_1, x_2, x_3)}{ds_1}$$

where along the x_1 axis it has been moved from x_1 by dx_1 to $x_1 + dx_1$ in finding the difference in φ for the numerator. The corresponding *distance* ds_1 in the denominator is from $ds_i = g_i\, dx_i$ taken as $g_1\, dx_1$ (see above) and substituted in the expression

$$(grad\varphi)_1 = \lim_{ds_1\to 0} \frac{\varphi(x_1 + dx_1, x_2, x_3) - \varphi(x_1, x_2, x_3)}{ds_1} = \lim_{g_1 dx_1\to 0} \frac{\varphi(x_1 + dx_1, x_2, x_3) - \varphi(x_1, x_2, x_3)}{g_1 dx_1}$$

In the limit, this becomes the partial derivative with respect to the variable x_1

$$(grad\varphi)_1 = \frac{1}{g_1}\frac{\partial\varphi(x_1, x_2, x_3)}{\partial x_1},\ (grad\varphi)_2 = \frac{1}{g_2}\frac{\partial\varphi(x_1, x_2, x_3)}{\partial x_2} \text{ and } (grad\varphi)_3 = \frac{1}{g_3}\frac{\partial\varphi(x_1, x_2, x_3)}{\partial x_3}.$$

Recall that $g_i^2 \equiv \left(\frac{\partial \mathbf{r}}{\partial x_i}\right)^2$, and the unit vectors in general coordinate systems vary with the coordinates from point to point. The expressions for $(grad\varphi)_i$ represent the projection of $grad\varphi$ on the varying unit vectors mentioned.

The divergence in general coordinates

The general definition for the divergence is independent of the coordinate system

$$div\ \mathbf{v} = \lim_{\Delta V\to 0} \frac{\oint_A \mathbf{v} d\mathbf{A}}{\Delta V}$$

It is independent of the form of the surface area. One obtains

$$\oint_A \mathbf{v} d\mathbf{A} = \left(\frac{\partial g_2 g_3 v_1}{\partial x_1} + \frac{\partial g_1 g_3 v_2}{\partial x_2} + \frac{\partial g_1 g_2 v_3}{\partial x_3}\right) dx_1 dx_2 dx_3$$

The volume ΔV involving the (infinitesimal geometrical) distances ds_i is given as

$$\Delta V \approx ds_1 ds_2 ds_3 = g_1 dx_1\ g_2 dx_2\ g_3 dx_3$$

The expression for the divergence becomes

$$div\ \mathbf{v} = \lim_{\Delta V\to 0} \frac{\oint_A \mathbf{v} d\mathbf{A}}{\Delta V} = \lim_{\Delta V\to 0} \frac{1}{g_1 dx_1\ g_2 dx_2\ g_3 dx_3}\left(\frac{\partial g_2 g_3 v_1}{\partial x_1} + \frac{\partial g_1 g_3 v_2}{\partial x_2} + \frac{\partial g_1 g_2 v_3}{\partial x_3}\right) dx_1 dx_2 dx_3$$

$$div\ \mathbf{v} = \frac{1}{g_1 g_2 g_3}\left[\frac{\partial}{\partial x_1}(g_2 g_3 v_1) + \frac{\partial}{\partial x_2}(g_1 g_3 v_2) + \frac{\partial}{\partial x_3}(g_1 g_2 v_3)\right]$$

The curl in general coordinates

The general definition for *curl* is independent of the coordinate system

$$(curl\ \mathbf{v})_n = \lim_{\Delta A \to 0} \frac{1}{\Delta A} \oint_L \mathbf{v} d\mathbf{l}$$

After some routine work, it may be written

$$(curl\ \mathbf{v})_1 = \frac{1}{g_2 g_3}\left[+\frac{\partial}{\partial x_2}(g_3 v_3) - \frac{\partial}{\partial x_3}(g_2 v_2)\right]$$

By cyclic substitution of the indices, the remaining components are given as

$$(curl\ \mathbf{v})_2 = \frac{1}{g_3 g_1}\left[+\frac{\partial}{\partial x_3}(g_1 v_1) - \frac{\partial}{\partial x_1}(g_3 v_3)\right] \qquad (curl\ \mathbf{v})_3 = \frac{1}{g_1 g_2}\left[+\frac{\partial}{\partial x_1}(g_2 v_2) - \frac{\partial}{\partial x_2}(g_1 v_1)\right]$$

The Laplace operator

By definition $div\ grad \equiv \Delta$

Recalling from previously

$$div\ \mathbf{v} = \frac{1}{g_1 g_2 g_3}\left[\frac{\partial}{\partial x_1}(g_2 g_3 v_1) + \frac{\partial}{\partial x_2}(g_1 g_3 v_2) + \frac{\partial}{\partial x_3}(g_1 g_2 v_3)\right]$$

$$(grad\varphi)_i = \frac{1}{g_i}\frac{\partial \varphi(x_1, x_2, x_3)}{\partial x_i} \qquad \text{from which} \qquad v_i = \frac{1}{g_i}\frac{\partial \varphi}{\partial x_i}$$

On substituting

$$\Delta\varphi = \frac{1}{g_1 g_2 g_3}\left[\frac{\partial}{\partial x_1}(g_2 g_3 \frac{1}{g_1}\frac{\partial \varphi}{\partial x_1}) + \frac{\partial}{\partial x_2}(g_1 g_3 \frac{1}{g_2}\frac{\partial \varphi}{\partial x_2}) + \frac{\partial}{\partial x_3}(g_1 g_2 \frac{1}{g_3}\frac{\partial \varphi}{\partial x_3})\right]$$

Common orthogonal coordinates

Cartesian Coordinates $\mathbf{x} = (x_1, x_2, x_3)$, $\mathbf{r} = (x, y, z)$		
x_1	x	Identity mapping between $\mathbf{x} = (x_1, x_2, x_3)$ and $\mathbf{r} = (x, y, z)$
x_2	y	
x_3	z	
ds_1	$ds_1 = dx$ Detail: $\frac{\partial \mathbf{r}}{\partial r} = (1,0,0)$ $\left(\frac{\partial \mathbf{r}}{\partial r}\right)^2 = 1^2 + 0^2 + 0^2 = 1^2$	$ds_i = g_i dx_i$, $i = 1,2,3$ Move in the x direction by dx, the elementary distance is $ds_1 = dx$ It follows that $g_1 = 1$ Or from $g_i^2 \equiv \left(\frac{\partial \mathbf{r}}{\partial x_i}\right)^2$ Same result for g_1
ds_2	$ds_2 = dy$	

ds_3	$ds_3 = dz$	
g_1	$g_1 = 1$	From $ds_1 = g_1 dx_1 \equiv dx$
g_2	$g_2 = 1$	
g_3	$g_3 = 1$	
Metric[1]	Infinitesimal Displacement: $ds = \sqrt{dx^2 + dy^2 + dz^2}$ Infinitesimal surfaces: $dA_{x\ const} = dy \cdot dz$ $dA_{y\ const} = dz \cdot dx$ $dA_{z\ const} = dx \cdot dy$ Infinitesimal volume: $dV = dx \cdot dy \cdot dz$	Elementary Geometric Distance is $\sqrt{\sum_{i=1}^{3} ds_i^2}$
∇	$\nabla \equiv \begin{pmatrix} \frac{\partial}{\partial x} \\ \frac{\partial}{\partial y} \\ \frac{\partial}{\partial z} \end{pmatrix}$	Nabla operator Cartesian components
Δ	$\Delta \equiv \frac{\partial^2}{\partial x^2} + \frac{\partial^2}{\partial y^2} + \frac{\partial^2}{\partial z^2}$	
Gradient	$\nabla\Phi = \begin{pmatrix} \frac{\partial}{\partial x}\Phi \\ \frac{\partial}{\partial y}\Phi \\ \frac{\partial}{\partial z}\Phi \end{pmatrix}$	Scalar Φ
Divergence	$div\,\mathbf{A} \equiv \nabla \cdot \mathbf{A} = \frac{\partial}{\partial x}A_x + \frac{\partial}{\partial y}A_y + \frac{\partial}{\partial z}A_z$	Scalar product; Also, Cf. above:
Curl	$curl\,\mathbf{A} \equiv \nabla \times \mathbf{A} = \begin{pmatrix} \frac{\partial}{\partial y}A_z - \frac{\partial}{\partial z}A_y \\ -\frac{\partial}{\partial x}A_z + \frac{\partial}{\partial z}A_x \\ \frac{\partial}{\partial x}A_y - \frac{\partial}{\partial y}A_x \end{pmatrix}$	Cross product; Also, Cf. above.

[1] The term metric is common usage among physicists who actually mean the norm.

Cylindrical Coordinates

$\mathbf{x} = (r, \varphi, z)$, $\mathbf{r} = (r\cos\varphi, r\sin\varphi, z)$

Cartesian ↔ Cylindrical

	$x = r\cos\varphi$ $y = r\sin\varphi$ $z = z$	$r = \sqrt{x^2 + y^2}$ $\varphi = \arctan\frac{y}{x}$ $z = z$
x_1	r	$0 \le r < \infty$
x_2	φ	$0 \le \varphi < 2\pi$
x_3	z	$0 \le z \le \infty$
ds_1	$ds_1 = dr$ Detail: $\frac{\partial \mathbf{r}}{\partial r} = (\cos\varphi, \sin\varphi, 0)$ $\left(\frac{\partial \mathbf{r}}{\partial r}\right)^2 = \cos^2\varphi + \sin^2\varphi + 0^2 = 1$	$ds_i = g_i\,dx_i$, $i = 1,2,3$ Move in the r direction by dr , the elementary distance is $ds_1 = dr$ It follows that $g_1 = 1$; Or from $g_i^2 \equiv \left(\frac{\partial \mathbf{r}}{\partial x_i}\right)^2$ Same result for g_1
ds_2	$ds_2 = r \cdot d\varphi$ $\frac{\partial \mathbf{r}}{\partial \varphi} = (-r\sin\varphi, r\cos\varphi, 0)$ $\left(\frac{\partial \mathbf{r}}{\partial \varphi}\right)^2 = r^2\sin^2\varphi + r^2\cos^2\varphi + 0^2 = r^2$	Move in the φ direction by $d\varphi$, the elementary distance is $ds_2 = r \cdot d\varphi$
ds_3	$ds_3 = dz$ Detail: $\frac{\partial \mathbf{r}}{\partial z} = (0,0,1)$ $\left(\frac{\partial \mathbf{r}}{\partial z}\right)^2 = 0^2 + 0^2 + 1^2 = 1$	Move in the z direction by dz , the elementary distance is $ds_3 = dz$
g_1	$g_1 = 1$	From $ds_1 = g_1\,dx_1 \equiv dr$
g_2	$g_2 = r$	
g_3	$g_3 = 1$	
Metric	Infinitesimal Displacement: $ds = \sqrt{dr^2 + r^2 d\varphi^2 + dz^2}$ Infinitesimal surfaces: $dA_{r\ const} = rd\varphi \cdot dz$ $dA_{\varphi\ const} = dr \cdot dz$ $dA_{z\ const} = rd\varphi \cdot dr$ Infinitesimal volume: $dV = dr \cdot rd\varphi \cdot dz$	Elementary Geometric Distance is $\sqrt{\sum_{i=1}^{3} ds_i^2}$

Cylinder ←Cartes	Cylindrical coordinates in terms of Cartesian coordinates: $\begin{pmatrix} A_r \\ A_\varphi \\ A_z \end{pmatrix} = \begin{pmatrix} \cos\varphi & \sin\varphi & 0 \\ -\sin\varphi & \cos\varphi & 0 \\ 0 & 0 & 1 \end{pmatrix} \cdot \begin{pmatrix} A_x \\ A_y \\ A_z \end{pmatrix}$ Conversely, using $\cos\varphi = x/r$ and $\sin\varphi = y/r$: $\begin{pmatrix} A_x \\ A_y \\ A_z \end{pmatrix} = \begin{pmatrix} \cos\varphi & -\sin\varphi & 0 \\ \sin\varphi & \cos\varphi & 0 \\ 0 & 0 & 1 \end{pmatrix} \cdot \begin{pmatrix} A_r \\ A_\varphi \\ A_z \end{pmatrix} =$ $\begin{pmatrix} A_x \\ A_y \\ A_z \end{pmatrix} = \begin{pmatrix} \frac{x}{\sqrt{x^2+y^2}} & \frac{-y}{\sqrt{x^2+y^2}} & 0 \\ \frac{y}{\sqrt{x^2+y^2}} & \frac{x}{\sqrt{x^2+y^2}} & 0 \\ 0 & 0 & 1 \end{pmatrix} \cdot \begin{pmatrix} A_r \\ A_\varphi \\ A_z \end{pmatrix}$	
∇	$\nabla \equiv \begin{pmatrix} \frac{\partial}{\partial r} \\ \frac{1}{r} \cdot \frac{\partial}{\partial \varphi} \\ \frac{\partial}{\partial z} \end{pmatrix}$	Nabla operator Cylindrical components
Δ	$\Delta \equiv \frac{1}{r}\left[\frac{\partial}{\partial r}(r\frac{\partial}{\partial r}) + \frac{\partial}{\partial \varphi}(\frac{1}{r}\frac{\partial}{\partial \varphi}) + \frac{\partial}{\partial z}(r\frac{\partial}{\partial z})\right]$	
Gradient	$\nabla\Phi = \begin{pmatrix} \frac{\partial}{\partial r}\Phi \\ \frac{1}{r} \cdot \frac{\partial}{\partial \varphi}\Phi \\ \frac{\partial}{\partial z}\Phi \end{pmatrix}$	Scalar Φ
Divergence	$div\,\mathbf{A} = \frac{1}{r}\frac{\partial}{\partial r}(rA_r) + \frac{1}{r}\frac{\partial}{\partial \varphi}A_\varphi + \frac{\partial}{\partial z}A_z$	Cf. above.

Curl	$curl\,\mathbf{A} = \begin{pmatrix} \frac{1}{r}\frac{\partial}{\partial\varphi}A_z - \frac{\partial}{\partial z}A_\varphi \\ -\frac{\partial}{\partial r}A_z + \frac{\partial}{\partial z}A_r \\ \frac{1}{r}\frac{\partial}{\partial r}(rA_\varphi) - \frac{1}{r}\frac{\partial}{\partial\varphi}A_r \end{pmatrix}$	Cf. above.

Spherical Coordinates

$$\mathbf{x} = (r, \theta, \varphi),\ \mathbf{r} = (r\sin\theta\cos\varphi, r\sin\theta\sin\varphi, r\cos\theta)$$

Cartesian ↔ Spherical

$$x = r\sin\theta\cos\varphi \qquad r = \sqrt{x^2 + y^2 + z^2}$$

$$y = r\sin\theta\sin\varphi \qquad \theta = \arccos\frac{z}{\sqrt{x^2 + y^2 + z^2}}$$

$$z = r\cos\theta \qquad \varphi = \arctan\frac{y}{x}$$

Cylindrical ↔ Spherical

$$\rho = r\sin\theta \qquad r = \sqrt{\rho^2 + z^2} \qquad \rho^2 = x^2 + y^2$$

$$\varphi = \varphi \qquad \theta = \arctan\frac{\rho}{z} = \arccos\frac{z}{\sqrt{\rho^2 + z^2}}$$

$$z = r\cos\theta \qquad \varphi = \varphi$$

x_1	r	$0 \le r < \infty$
x_2	θ	$0 \le \theta < \pi$
x_3	φ	$0 \le \varphi < 2\pi$
ds_1	$ds_1 = dr$	$ds_i = g_i\,dx_i$, $i = 1,2,3$ Move in the r direction by dr, the elementary distance is $ds_1 = dr$ It follows that $g_1 = 1$ Or from $g_i^2 \equiv \left(\frac{\partial\mathbf{r}}{\partial x_i}\right)^2$ Same result for g_1
ds_2	$ds_2 = r \cdot d\theta$	Move in the θ direction by $d\theta$, the elementary distance is $ds_2 = r \cdot d\theta$
ds_3	$ds_3 = r\sin\theta \cdot d\varphi$	Move in the φ direction by $d\varphi$, the elementary distance is

		$ds_3 = r\sin\theta \cdot d\varphi$
g_1	$g_1 = 1$	From $ds_1 = g_1\, dx_1 \equiv dr$
g_2	$g_2 = r$	
g_3	$g_3 = r\sin\theta$	
Metric	Infinitesimal Displacement: $ds = \sqrt{dr^2 + r^2 d\theta^2 + (r\sin\theta\, d\varphi)^2}$ Infinitesimal surfaces: $dA_{\ r\,const} = rd\theta \cdot r\sin\theta\, d\varphi$ (Known as space angle) $dA_{\ \varphi\,const} = dr \cdot rd\theta$ $dA_{\ \theta\,const} = dr \cdot r\sin\theta\, d\varphi$ Infinitesimal volume: $dV = dr \cdot rd\theta \cdot r\sin\theta\, d\varphi$	Elementary Geometric Distance is $\sqrt{\sum_{i=1}^{3} ds_i^2}$
Sphere ←Cartesian	Spherical coordinates in terms of Cartesian coordinates: $\begin{pmatrix} A_r \\ A_\theta \\ A_\varphi \end{pmatrix} = \begin{pmatrix} \cos\varphi\sin\theta & \sin\varphi\sin\theta & \cos\theta \\ \cos\varphi\cos\theta & \sin\varphi\cos\theta & -\sin\theta \\ -\sin\varphi & \cos\varphi & 0 \end{pmatrix} \cdot \begin{pmatrix} A_x \\ A_y \\ A_z \end{pmatrix}$ Conversely, $\begin{pmatrix} A_x \\ A_y \\ A_z \end{pmatrix} = \begin{pmatrix} \frac{x}{\sqrt{x^2+y^2+z^2}} & \frac{xz/\sqrt{x^2+y^2}}{\sqrt{x^2+y^2+z^2}} & \frac{-y}{\sqrt{x^2+y^2}} \\ \frac{y}{\sqrt{x^2+y^2+z^2}} & \frac{yz/\sqrt{x^2+y^2}}{\sqrt{x^2+y^2+z^2}} & \frac{x}{\sqrt{x^2+y^2}} \\ \frac{z}{\sqrt{x^2+y^2+z^2}} & \frac{-\sqrt{x^2+y^2}}{\sqrt{x^2+y^2+z^2}} & 0 \end{pmatrix} \cdot \begin{pmatrix} A_r \\ A_\theta \\ A_\varphi \end{pmatrix}$	
Cylinder ←Sphere	Cylindrical coordinates in terms of Spherical coordinates $\begin{pmatrix} A_\rho \\ A_\varphi \\ A_z \end{pmatrix} = \begin{pmatrix} \frac{\rho}{\sqrt{\rho^2+z^2}} & \frac{z}{\sqrt{\rho^2+z^2}} & 0 \\ 0 & 0 & 1 \\ \frac{z}{\sqrt{\rho^2+z^2}} & \frac{-\rho}{\sqrt{\rho^2+z^2}} & 0 \end{pmatrix} \cdot \begin{pmatrix} A_r \\ A_\theta \\ A_\varphi \end{pmatrix}$ Conversely, $\begin{pmatrix} A_r \\ A_\theta \\ A_\varphi \end{pmatrix} = \begin{pmatrix} \sin\theta & 0 & \cos\theta \\ \cos\theta & 0 & -\sin\theta \\ 0 & 1 & 0 \end{pmatrix} \cdot \begin{pmatrix} A_\rho \\ A_\varphi \\ A_z \end{pmatrix}$	Note: $\rho^2 = x^2 + y^2$

∇	$\nabla \equiv \begin{pmatrix} \frac{\partial}{\partial r} \\ \frac{1}{r} \cdot \frac{\partial}{\partial \theta} \\ \frac{1}{r \sin\theta} \cdot \frac{\partial}{\partial \varphi} \end{pmatrix}$	Nabla operator Spherical components
Δ	$\Delta \equiv \frac{1}{r^2} \frac{\partial}{\partial r}(r^2 \frac{\partial}{\partial r}) + \frac{1}{r^2 \sin\theta} \frac{\partial}{\partial \theta}(\sin\theta \frac{\partial}{\partial \theta}) + \frac{1}{r^2 \sin^2\theta} \frac{\partial^2}{\partial \varphi^2}$	
Gradient	$\nabla \Phi = \begin{pmatrix} \frac{\partial}{\partial r}\Phi \\ \frac{1}{r} \cdot \frac{\partial}{\partial \theta}\Phi \\ \frac{1}{r \sin\theta} \cdot \frac{\partial}{\partial \varphi}\Phi \end{pmatrix}$	Scalar Φ
Divergence	$div\,\mathbf{A} = \frac{1}{r^2} \frac{\partial}{\partial r}(r^2 A_r) + \frac{1}{r \sin\theta}\left[\frac{\partial}{\partial \theta}(\sin\theta\, A_\theta) + \frac{\partial}{\partial \varphi} A_\varphi\right]$	Cf. above;
Curl	$curl\,\mathbf{A} = \begin{pmatrix} \frac{1}{r \sin\theta} \frac{\partial}{\partial \theta}(\sin\theta\, A_\varphi) - \frac{1}{r \sin\theta} \frac{\partial}{\partial \varphi} A_\theta \\ \frac{1}{r \sin\theta} \frac{\partial}{\partial \varphi} A_r - \frac{1}{r} \frac{\partial}{\partial r}(r A_\varphi) \\ \frac{1}{r} \frac{\partial}{\partial r}(r A_\theta) - \frac{1}{r} \frac{\partial}{\partial \theta} A_r \end{pmatrix}$	Cf. above;

Detail for Δ

Recalling that

$$\Delta\varphi = \frac{1}{g_1 g_2 g_3}\left[\frac{\partial}{\partial x_1}(g_2 g_3 \frac{1}{g_1} \frac{\partial \varphi}{\partial x_1}) + \frac{\partial}{\partial x_2}(g_1 g_3 \frac{1}{g_2} \frac{\partial \varphi}{\partial x_2}) + \frac{\partial}{\partial x_3}(g_1 g_2 \frac{1}{g_3} \frac{\partial \varphi}{\partial x_3})\right]$$

On substituting

$$\Delta = \frac{1}{r^2 \sin\theta}\left[\frac{\partial}{\partial r}(r^2 \sin\theta \frac{\partial}{\partial r}) + \frac{\partial}{\partial \theta}(\sin\theta \frac{\partial}{\partial \theta}) + \frac{\partial}{\partial \varphi}(\frac{1}{\sin\theta} \frac{\partial}{\partial \varphi})\right]$$

Pulling the terms unaffected by the partial differentiations out of the brackets

$$\Delta = \frac{1}{r^2 \sin\theta}\left[\sin\theta \frac{\partial}{\partial r}(r^2 \frac{\partial}{\partial r}) + \frac{\partial}{\partial \theta}(\sin\theta \frac{\partial}{\partial \theta}) + \frac{1}{\sin\theta} \frac{\partial}{\partial \varphi}(\frac{\partial}{\partial \varphi})\right]$$

Rearranging

$$\Delta \equiv \frac{1}{r^2} \frac{\partial}{\partial r}(r^2 \frac{\partial}{\partial r}) + \frac{1}{r^2 \sin\theta} \frac{\partial}{\partial \theta}(\sin\theta \frac{\partial}{\partial \theta}) + \frac{1}{r^2 \sin^2\theta} \frac{\partial^2}{\partial \varphi^2}$$

Field Equations

TM

It is assumed metal conducting element with given current distribution $\mathbf{J}$ embedded in medium mcharacterized by conductivity σ, dielectric constant ε and permeability μ

I.) $curl\ \mathbf{H} = \mathbf{J} + (\sigma + j\omega\varepsilon)\,\mathbf{E}$ III.) $div\ \mathbf{H} = 0$

II.) $curl\ \mathbf{E} = -j\omega\mu\ \mathbf{H}$ IV.) $div\ \mathbf{E} = \dfrac{\rho}{\varepsilon}$

Recalling that $\mathbf{A} = \frac{1}{k} curl\mathbf{B} \leftrightarrow \mathbf{B} = \frac{1}{k} curl\mathbf{A}$, it is important to see from II.) that an Ansatz such as $\mathbf{E} = -j\omega\mu\ curl\mathbf{\Pi}_m$ may be made. Likewise from I.) the Ansatz would read $\mathbf{H} = (\sigma + j\omega\varepsilon) curl\mathbf{\Pi}_e$ despite $\mathbf{J} \neq 0$

Starting from III.), by the identity $div\, curl \equiv 0$, $\mathbf{H} = (\sigma + j\omega\varepsilon) curl\mathbf{\Pi}_e$ satisfies III.)
$\mathbf{\Pi}_e$ and $\mathbf{H}$ are perpendicular to each other. Thus assuming that the former had only one component in a distinct direction, for example the z direction, in which case the latter could not possess a component in that same direction for the reason just mentioned. The magnetic field would be termed TM.
Substituting into II.) $curl\ \mathbf{E} = -j\omega\mu(\sigma + j\omega\varepsilon) curl\mathbf{\Pi}_e$ thus $curl\ \mathbf{E} = k^2 curl\mathbf{\Pi}_e$ where $k^2 = -j\omega\mu(\sigma + j\omega\varepsilon)$, followed by grouping all terms under the *curl* operator i.e. $curl\ (\mathbf{E} - k^2\mathbf{\Pi}_e) = 0$, followed by using another identity $curl\, grad \equiv 0$ for another Ansatz namely $\mathbf{E} - k^2\mathbf{\Pi}_e = +grad\phi$. Accordingly $\mathbf{E} = k^2\mathbf{\Pi}_e + grad\phi$ where ϕ is yet to be *chosen* such that the expressions would become simplest to handle.
With $\mathbf{H}$ and $\mathbf{E}$ now available, substituting back into I.) to have $curl\ (\sigma + j\omega\varepsilon) curl\mathbf{\Pi}_e = \mathbf{J} + (\sigma + j\omega\varepsilon)\{k^2\mathbf{\Pi}_e + grad\phi\}$, i.e. $curl\ curl\mathbf{\Pi}_e = \frac{\mathbf{J}}{(\sigma + j\omega\varepsilon)} + (k^2\mathbf{\Pi}_e + grad\phi)$ which represents an equation for $\mathbf{\Pi}_e$

IV.) becomes $div\ (k^2\mathbf{\Pi}_e + grad\phi) = \dfrac{\rho}{\varepsilon}$, which on using the shorthand $div\, grad \equiv \Delta$ in $div\, k^2\mathbf{\Pi}_e + div\ grad\phi = \dfrac{\rho}{\varepsilon}$ reads $k^2 div\,\mathbf{\Pi}_e + \Delta\phi = \dfrac{\rho}{\varepsilon}$. It thus makes sense to be *choosing* $div\,\mathbf{\Pi}_e = \phi$ so as to get to the familiar scalar wave equation $\Delta\phi + k^2\phi = \dfrac{\rho}{\varepsilon}$ for space charge density $\rho \neq 0$.

Homogeneous cases would result for $\rho = 0$. One or two words should be mentioned about the wave eqs., Cf. Footnote[2]

---

[2] Returning to $k^2 = -j\omega\mu(\sigma + j\omega\varepsilon)$, so as to rewrite $\Delta\phi - j\omega\mu(\sigma + j\omega\varepsilon)\phi = \frac{\rho}{\varepsilon}$ as $\Delta\phi - \mu\varepsilon\frac{\partial^2}{\partial t^2}\phi = \frac{\rho}{\varepsilon}$, a solution for which may be given $\phi = \frac{1}{4\pi}\int \frac{-\Delta\phi}{r}dV$ in the static case of $\frac{\partial}{\partial t} \equiv 0$, the static case being considered for ease of discussion. Choosing $\mathbf{E} - k^2\mathbf{\Pi}_e = -grad\varphi$ would have led to $\Delta\varphi - \mu\varepsilon\frac{\partial^2}{\partial t^2}\varphi = \frac{-\rho}{\varepsilon}$, with solution $\varphi = \frac{1}{4\pi}\int \frac{-\Delta\varphi}{r}dV$ for $\frac{\partial}{\partial t} \equiv 0$. Obviously $\phi = -\varphi$.

Returning to $curl\ curl\mathbf{\Pi}_e = \frac{\mathbf{J}}{(\sigma + j\omega\varepsilon)} + (k^2\mathbf{\Pi}_e + grad\phi)$ and using $\Delta \equiv grad\ div - curl\ curl$ to now write $grad\ div\mathbf{\Pi}_e - \Delta\mathbf{\Pi}_e = \frac{\mathbf{J}}{(\sigma + j\omega\varepsilon)} + (k^2\mathbf{\Pi}_e + grad\ div\mathbf{\Pi}_e)$, which becomes $-\Delta\mathbf{\Pi}_e = \frac{\mathbf{J}}{(\sigma + j\omega\varepsilon)} + k^2\mathbf{\Pi}_e$ that is the vector wave equation for $\mathbf{\Pi}_e$, that is identical to the scalar one, namely $\Delta\mathbf{\Pi}_e + k^2\mathbf{\Pi}_e = \frac{-\mathbf{J}}{(\sigma + j\omega\varepsilon)}$. Again for ease of discussion let $\sigma \to 0$. Substituting $\mathbf{A} = \varepsilon j\omega\mathbf{\Pi}_e$ to have $\Delta\mathbf{A} + k^2\mathbf{A} = -\mathbf{J}$ a solution for which may be written $\mathbf{A} = \frac{1}{4\pi}e^{j\omega t}\int \frac{\mathbf{J}e^{-jkr}}{r}dV$ important to stationary cases and linear antennas. Notice the generic argument of $e^{j\omega\left(t - \frac{r}{c}\right)}$.

One more word should be spent on $\frac{\mathbf{J}}{j\omega\varepsilon}$. Writing $\frac{\mathbf{A}}{j\omega\varepsilon} = \frac{1}{4\pi}e^{j\omega t}\int \frac{\frac{\mathbf{J}}{j\omega\varepsilon}e^{-jkr}}{r}dV$ i.e. $\mathbf{\Pi}_e = \frac{1}{4\pi\varepsilon}e^{j\omega t}\int \frac{\mathbf{f}e^{-jkr}}{r}dV$ where $\frac{\mathbf{J}}{j\omega} = \mathbf{f}$ i.e. $\mathbf{J} = \frac{\partial}{\partial t}\mathbf{f}$.

Independently from the lines just shown, how did the substitution $\mathbf{J} = \frac{\partial}{\partial t}\mathbf{f}$ come about? Consider

$$div_Q(\xi_i\mathbf{J}) = grad_Q\xi_i\mathbf{J} + \xi_i div_Q\mathbf{J}$$
$$div_Q(\xi_i\mathbf{J}) - \xi_i div_Q\mathbf{J} = grad_Q\xi_i\mathbf{J},$$

but $grad_Q\xi_i\mathbf{J} = J_i$, thus

$$\int J_i dV_Q = \int div_Q(\xi_i\mathbf{J})dV_Q - \int \xi_i div_Q\mathbf{J}dV_Q$$

By the Gauß theorem, where knowing that the surface integral is zero

$$\int J_i dV_Q = \oint \xi_i\mathbf{J}dA_Q - \int \xi_i div_Q\mathbf{J}dV_Q$$

$$\int J_i dV_Q = -\int \xi_i div_Q\mathbf{J}dV_Q$$

Using $div\mathbf{J} = -\frac{\partial}{\partial t}\rho$

$$\int J_i dV_Q = -\int \xi_i(-\frac{\partial}{\partial t}\rho)dV_Q = +j\omega\int \xi_i\rho dV_Q$$

Writing $p_i = \int \xi_i\rho dV_Q \equiv \int f_i dV_Q$ that is $\mathbf{p} = \int \mathbf{f}dV_Q$ to have

In summary, $\mathbf{H}$ and $\mathbf{E}$ would be computed as shown, if $\mathbf{\Pi}_e$ has been found from $curl\ curl\mathbf{\Pi}_e = \frac{\mathbf{J}}{(\sigma + j\omega\varepsilon)} + (k^2\mathbf{\Pi}_e + grad\phi)$. Notice that $\mathbf{E} = curl\ curl\mathbf{\Pi}_e$ applies *only* for where there is no conducting current flowing $\mathbf{J} = 0$.

TE

I.)	$curl\ \mathbf{H} = \mathbf{J} + (\sigma + j\omega\varepsilon)\,\mathbf{E}$	III.)	$div\ \mathbf{H} = 0$
II.)	$curl\ \mathbf{E} = -j\omega\mu\ \mathbf{H}$	IV.)	$div\ \mathbf{E} = 0$

Starting from IV.) that is modified to zero as shown for the Ansatz $\mathbf{E} = -j\omega\mu\ curl\mathbf{\Pi}_m$ to be usable. Substituting into I.) $curl\ \mathbf{H} = \mathbf{J} + (\sigma + j\omega\varepsilon)\{-j\omega\mu\ curl\mathbf{\Pi}_m\}$ thus $curl\ \mathbf{H} = \mathbf{J} + k^2 curl\mathbf{\Pi}_m$, which indicates another modification $\mathbf{J} = 0$ is required so as to be able to group all terms under the *curl* operator, $curl\ (\mathbf{H} - k^2\mathbf{\Pi}_m) = 0$. Ansatz $\mathbf{H} - k^2\mathbf{\Pi}_m = +grad\Phi$. Accordingly $\mathbf{H} = k^2\mathbf{\Pi}_m + grad\Phi$ where Φ is to be *chosen* such that the expressions would become simplest to handle.
With $\mathbf{H}$ and $\mathbf{E}$ now available, substituting back into II.) to have $curl\ (-j\omega\mu\ curl\mathbf{\Pi}_m) = -j\omega\mu(k^2\mathbf{\Pi}_m + grad\Phi)$, i.e. $curl\ curl\mathbf{\Pi}_m = (k^2\mathbf{\Pi}_m + grad\Phi)$, which represents an equation for $\mathbf{\Pi}_m$, which is identical to the above $curl\ curl\mathbf{\Pi}_e = \frac{\mathbf{J}}{(\sigma + j\omega\varepsilon)} + (k^2\mathbf{\Pi}_e + grad\varphi)$ for $\mathbf{J} = 0$. Also notice that $curl\ curl\mathbf{\Pi}_m = \mathbf{H}$
III.) is automatically satisfied from $curl\ curl\mathbf{\Pi}_m = \mathbf{H}$. Furthermore the homogeneous scalar wave equation $\Delta\Phi + k^2\Phi = 0$ results from *choosing* $div\,\mathbf{\Pi}_m = \Phi$

In summary, $\mathbf{H}$ and $\mathbf{E}$ would be computed as shown, if $\mathbf{\Pi}_m$ has been found from $curl\ curl\mathbf{\Pi}_m = (k^2\mathbf{\Pi}_m + grad\Phi)$. Notice that $\mathbf{H} = curl\ curl\mathbf{\Pi}_m$.

§§§

$\int J_i dV_Q = j\omega p_i$, is same as $\int \mathbf{J} dV_Q = j\omega\mathbf{p}$

(The LHS of the latter may be written $\mathbf{l}I$, boldface l times current involved in the expression for the elementary dipole antenna, thus $\mathbf{p} = \mathbf{l}I / j\omega = \mathbf{l}q$ known as the dipole's moment)

Inserting $\mathbf{p} = \int \mathbf{f} dV_Q$ into $\int \mathbf{J} dV_Q = j\omega\mathbf{p}$ gets to $\int \mathbf{J} dV_Q = j\omega\mathbf{p} = j\omega\int \mathbf{f} dV_Q$

The substitution $\mathbf{J} = \frac{\partial}{\partial t}\mathbf{f}$ came about from there.

Spherical Coordinates

Procedure

The Maxwell Eqs. led to an intractable equation for the general case namely
$grad\ \phi - curl\ curl\ \mathbf{\Pi} + k^2\mathbf{\Pi} = 0$
where $\mathbf{\Pi}$ is the Hertz vector.

If $\mathbf{\Pi}$ were known, the field vectors would be computed from
$\mathbf{H} = (\sigma + j\varepsilon\omega)\ curl\ \mathbf{\Pi}$
$\mathbf{E} = k^2\ \mathbf{\Pi} + grad\ \phi$
or
$\mathbf{E} = -j\mu\omega\ curl\ \mathbf{\Pi}$
$\mathbf{H} = k^2\ \mathbf{\Pi} + grad\ \phi$

Simplifying Ansätze have been made. Usually, $\mathbf{\Pi}$ is assumed to have only one distinct non-zero component. Also in accommodating some symmetric geometry, such as that of a concentric cylinder or a sphere, the choice for ϕ is to alleviate the efforts as far as possible. The Ansätze led to well known differential equations of Physics, with known solutions for $\mathbf{\Pi}$. (This text is showing one or two examples for the latter mentioned.)

A list is compiled for the expressions for the fields, taking $J_{n+\frac{1}{2}}(kr)$ as example in the following

$\Pi_r(r,\theta,\varphi;t) = \sqrt{kr}Z_{n+\frac{1}{2}}(kr)P_n^m(\cos\theta)\mathrm{e}^{\pm jm\varphi}\mathrm{e}^{j\omega t}$

$k^2 = -j\omega\mu(\sigma + j\omega\varepsilon)$ involving frequency and material constants

TM

$$E_r = k^2\Pi_r + \frac{\partial^2\Pi_r}{\partial r^2} = \frac{n(n+1)}{r^2}\Pi_r \qquad H_r = 0$$

$$E_\theta = \frac{1}{r}\frac{\partial^2\Pi_r}{\partial r\partial\theta} \qquad H_\theta = \frac{\varepsilon j\omega}{r\sin\theta}\frac{\partial\Pi_r}{\partial\varphi}$$

$$E_\varphi = \frac{1}{r\sin\theta}\frac{\partial^2\Pi_r}{\partial r\partial\varphi} \qquad H_\varphi = -\frac{\varepsilon j\omega}{r}\frac{\partial\Pi_r}{\partial\theta}$$

TE

$$E_r = 0 \qquad H_r = k^2\Pi_r + \frac{\partial^2\Pi_r}{\partial r^2} = \frac{n(n+1)}{r^2}\Pi_r$$

$$E_\theta = -\frac{\mu j\omega}{r\sin\theta}\frac{\partial \Pi_r}{\partial \varphi} \qquad H_\theta = \frac{1}{r}\frac{\partial^2 \Pi_r}{\partial r \partial \theta}$$

$$E_\varphi = \frac{\mu j\omega}{r}\frac{\partial \Pi_r}{\partial \theta} \qquad H_\varphi = \frac{1}{r\sin\theta}\frac{\partial^2 \Pi_r}{\partial r \partial \varphi}$$

Proceeding for TM

$$E_r = \frac{n(n+1)}{r^2}\sqrt{kr}J_{n+\frac{1}{2}}(kr)P_n^m(\cos\theta)\mathrm{e}^{jm\varphi} \qquad = k^2\frac{n(n+1)}{(kr)^{\frac{3}{2}}}J_{n+\frac{1}{2}}(kr)P_n^m(\cos\theta)\mathrm{e}^{jm\varphi}$$

$$E_\theta = \frac{1}{r}\frac{\partial}{\partial r}\left(\sqrt{kr}J_{n+\frac{1}{2}}(kr)\right)\frac{\partial}{\partial\theta}P_n^m(\cos\theta)\mathrm{e}^{jm\varphi}$$

$$E_\varphi = \frac{jm}{r\sin\theta}\frac{\partial}{\partial r}\left(\sqrt{kr}J_{n+\frac{1}{2}}(kr)\right)P_n^m(\cos\theta)\mathrm{e}^{jm\varphi}$$

$$H_r = 0$$

$$H_\theta = \frac{\varepsilon j\omega}{r\sin\theta}jm\sqrt{kr}J_{n+\frac{1}{2}}(kr)P_n^m(\cos\theta)\mathrm{e}^{jm\varphi} \qquad = \frac{-\varepsilon\omega m}{r\sin\theta}\sqrt{kr}J_{n+\frac{1}{2}}(kr)P_n^m(\cos\theta)\mathrm{e}^{jm\varphi}$$

$$H_\varphi = \frac{-\varepsilon j\omega}{r}\sqrt{kr}J_{n+\frac{1}{2}}(kr)\frac{\partial}{\partial\theta}P_n^m(\cos\theta)\mathrm{e}^{jm\varphi}$$

Proceeding for TE

$$E_r = 0$$

$$E_\theta = \frac{-\mu j\omega}{r\sin\theta}jm\sqrt{kr}J_{n+\frac{1}{2}}(kr)P_n^m(\cos\theta)\mathrm{e}^{jm\varphi} \qquad = \frac{\mu\omega m}{r\sin\theta}\sqrt{kr}J_{n+\frac{1}{2}}(kr)P_n^m(\cos\theta)\mathrm{e}^{jm\varphi}$$

$$E_\varphi = \frac{\mu j\omega}{r}\sqrt{kr}J_{n+\frac{1}{2}}(kr)\frac{\partial}{\partial\theta}P_n^m(\cos\theta)\mathrm{e}^{jm\varphi}$$

$$H_r = \frac{n(n+1)}{r^2}\sqrt{kr}J_{n+\frac{1}{2}}(kr)P_n^m(\cos\theta)\mathrm{e}^{jm\varphi} \qquad = k^2\frac{n(n+1)}{(kr)^{\frac{3}{2}}}J_{n+\frac{1}{2}}(kr)P_n^m(\cos\theta)\mathrm{e}^{jm\varphi}$$

$$H_\theta = \frac{1}{r}\frac{\partial}{\partial r}\left(\sqrt{kr}J_{n+\frac{1}{2}}(kr)\right)\frac{\partial}{\partial\theta}P_n^m(\cos\theta)\mathrm{e}^{jm\varphi}$$

$$H_\varphi = \frac{jm}{r\sin\theta}\frac{\partial}{\partial r}\left(\sqrt{kr}J_{n+\frac{1}{2}}(kr)\right)P_n^m(\cos\theta)\mathrm{e}^{jm\varphi}$$

Boundary conditions

$$\Pi_r(r,\theta,\varphi;t) = \sqrt{kr}Z_{n+\frac{1}{2}}(kr)P_n^m(\cos\theta)\mathrm{e}^{\pm jm\varphi}$$

$$k^2 = -j\omega\mu(\sigma + j\omega\varepsilon)$$

Take $Z_{n+\frac{1}{2}}(kr)$ to be $J_{n+\frac{1}{2}}(kr)$ or $H^{(2)}_{n+\frac{1}{2}}(kr)$ to ensure finite limits at zero and zero at infinity, respectively.

For TM (Similar procedure for TE) (by convention $a^{\cdot}_{mn}$ is shorthand for $\sum_{n,m} a^{\cdot}_{mn}$)

$$E_r^i = a_{mn}^i k_i^2\frac{n(n+1)}{(k_i r)^{\frac{3}{2}}}J_{n+\frac{1}{2}}(k_i r)P_n^m(\cos\theta)\mathrm{e}^{jm\varphi}$$

$$E_\theta^i = a_{mn}^i \frac{1}{r}\frac{\partial}{\partial r}\left(\sqrt{k_i r} J_{n+\frac{1}{2}}(k_i r)\right)\frac{\partial}{\partial \theta} P_n^m(\cos\theta)\,\mathrm{e}^{jm\varphi}$$

$$E_\varphi^i = a_{mn}^i \frac{jm}{r\sin\theta}\frac{\partial}{\partial r}\left(\sqrt{k_i r} J_{n+\frac{1}{2}}(k_i r)\right) P_n^m(\cos\theta)\,\mathrm{e}^{jm\varphi}$$

$$H_r^i = 0$$

$$H_\theta^i = a_{mn}^i \frac{-\varepsilon_i \omega m}{r\sin\theta}\sqrt{k_i r} J_{n+\frac{1}{2}}(k_i r) P_n^m(\cos\theta)\,\mathrm{e}^{jm\varphi}$$

$$H_\varphi^i = a_{mn}^i \frac{-\varepsilon_i j\omega}{r}\sqrt{k_i r} J_{n+\frac{1}{2}}(k_i r)\frac{\partial}{\partial \theta} P_n^m(\cos\theta)\,\mathrm{e}^{jm\varphi}$$

The index i denotes “inside” or from within the sphere.

The index a below (instead of i) is a reminder for zero *at infinity* i.e. outside of the sphere, where $H_{n+\frac{1}{2}}^{(2)}(k_a r)$ would replace $J_{n+\frac{1}{2}}(k_i r)$ for an exactly same set of equations that apply for the space outside of the sphere.

At $r = r_0$, for the tangential components of the field vectors to be continuous, the boundary conditions read

$$E_\theta^i = E_\theta^a \qquad E_\varphi^i = E_\varphi^a$$

$$H_\theta^i = H_\theta^a \qquad H_\varphi^i = H_\varphi^a$$

which translate into (element-by-element equality as a sufficient condition for equality between the summations is assumed in the following, thus $a_{mn}^{\cdot}$ meaning here an element under consideration)

$$a_{mn}^i \frac{\partial}{\partial r}\left(\sqrt{k_i r} J_{n+\frac{1}{2}}(k_i r)\right)_{r=r_0} = a_{mn}^a \frac{\partial}{\partial r}\left(\sqrt{k_a r} H_{n+\frac{1}{2}}^{(2)}(k_a r)\right)_{r=r_0}$$

$$a_{mn}^i \frac{k_i^2}{\mu_i}\sqrt{k_i r_0} J_{n+\frac{1}{2}}(k_i r_0) = a_{mn}^a \frac{k_a^2}{\mu_a}\sqrt{k_a r_0} H_{n+\frac{1}{2}}^{(2)}(k_a r_0)$$

$$\begin{pmatrix} \frac{\partial}{\partial r}\left(\sqrt{k_i r} J_{n+\frac{1}{2}}(k_i r)\right)_{r=r_0} & -\frac{\partial}{\partial r}\left(\sqrt{k_a r} H_{n+\frac{1}{2}}^{(2)}(k_a r)\right)_{r=r_0} \\ \frac{k_i^2}{\mu_i}\sqrt{k_i r_0} J_{n+\frac{1}{2}}(k_i r_0) & -\frac{k_a^2}{\mu_a}\sqrt{k_a r_0} H_{n+\frac{1}{2}}^{(2)}(k_a r_0) \end{pmatrix}\begin{pmatrix} a_{mn}^i \\ a_{mn}^a \end{pmatrix} = \begin{pmatrix} 0 \\ 0 \end{pmatrix}$$

The determinant is to be zero for the eq. system to have any non-trivial solutions at all

$$-\frac{\partial}{\partial r}\left(\sqrt{k_i r} J_{n+\frac{1}{2}}(k_i r)\right)_{r=r_0}\frac{k_a^2}{\mu_a}\sqrt{k_a r_0} H_{n+\frac{1}{2}}^{(2)}(k_a r_0) + \frac{k_i^2}{\mu_i}\sqrt{k_i r_0} J_{n+\frac{1}{2}}(k_i r_0)\frac{\partial}{\partial r}\left(\sqrt{k_a r} H_{n+\frac{1}{2}}^{(2)}(k_a r)\right)_{r=r_0} = 0$$

$$\frac{\partial}{\partial r}\left(\sqrt{k_i r} J_{n+\frac{1}{2}}(k_i r)\right)_{r=r_0}\frac{k_a^2}{\mu_a}\sqrt{k_a r_0} H_{n+\frac{1}{2}}^{(2)}(k_a r_0) = \frac{k_i^2}{\mu_i}\sqrt{k_i r_0} J_{n+\frac{1}{2}}(k_i r_0)\frac{\partial}{\partial r}\left(\sqrt{k_a r} H_{n+\frac{1}{2}}^{(2)}(k_a r)\right)_{r=r_0}$$

$$\frac{\frac{\partial}{\partial r}\left(\sqrt{k_a r} H_{n+\frac{1}{2}}^{(2)}(k_a r)\right)_{r=r_0}}{\frac{\partial}{\partial r}\left(\sqrt{k_i r} J_{n+\frac{1}{2}}(k_i r)\right)_{r=r_0}} = \frac{\frac{k_a^2}{\mu_a}\sqrt{k_a r_0} H_{n+\frac{1}{2}}^{(2)}(k_a r_0)}{\frac{k_i^2}{\mu_i}\sqrt{k_i r_0} J_{n+\frac{1}{2}}(k_i r_0)}$$

and thus

$$\frac{\frac{\partial}{\partial r}\left(\sqrt{k_a r} H_{n+\frac{1}{2}}^{(2)}(k_a r)\right)_{r=r_0}}{\frac{\partial}{\partial r}\left(\sqrt{k_i r} J_{n+\frac{1}{2}}(k_i r)\right)_{r=r_0}} = \frac{k_a^{\frac{5}{2}}}{k_i^{\frac{5}{2}}}\frac{\mu_i}{\mu_a}\frac{H_{n+\frac{1}{2}}^{(2)}(k_a r_0)}{J_{n+\frac{1}{2}}(k_i r_0)}$$

The variable substitution $\rho = kr$, $\partial\rho = k\partial r$ so as to differentiate w.r.t. the arguments will have the factor $\frac{k_a}{k_i}$ as a left-over on the LHS, that contributes to the result such that in end effect only the ratio $\frac{k_a^{\frac{3}{2}}}{k_i^{\frac{3}{2}}}$ is involved. More importantly, ω is the only variable in the equation shown. The boundary conditions thus require the existence of discrete values for ω known as angular Eigenfrequencies that are typically complex. Should some turn out to be purely imaginary, they would represent damping only from forming $\mathrm{e}^{\pm j\omega t}$.

TE

$$E_r^i = 0$$

$$E_\theta^i = a_{mn}^i \frac{\mu_i \omega m}{r\sin\theta}\sqrt{k_i r}J_{n+\frac{1}{2}}(k_i r)P_n^m(\cos\theta)\mathrm{e}^{jm\varphi}$$

$$E_\varphi^i = a_{mn}^i \frac{\mu_i j\omega}{r}\sqrt{k_i r}J_{n+\frac{1}{2}}(k_i r)\frac{\partial}{\partial\theta}P_n^m(\cos\theta)\mathrm{e}^{jm\varphi}$$

$$H_r^i = a_{mn}^i k_i^2 \frac{n(n+1)}{(k_i r)^{\frac{3}{2}}}J_{n+\frac{1}{2}}(k_i r)P_n^m(\cos\theta)\mathrm{e}^{jm\varphi}$$

$$H_\theta^i = a_{mn}^i \frac{1}{r}\frac{\partial}{\partial r}\left(\sqrt{k_i r}J_{n+\frac{1}{2}}(k_i r)\right)\frac{\partial}{\partial\theta}P_n^m(\cos\theta)\mathrm{e}^{jm\varphi}$$

$$H_\varphi^i = a_{mn}^i \frac{jm}{r\sin\theta}\frac{\partial}{\partial r}\left(\sqrt{k_i r}J_{n+\frac{1}{2}}(k_i r)\right)P_n^m(\cos\theta)\mathrm{e}^{jm\varphi}$$

$E_\theta^i = E_\theta^a$ gives at $r = r_0$ (a_{mn} meaning here an element under consideration)

$$a_{mn}^i \frac{\mu_i \omega m}{r\sin\theta}\sqrt{k_i r_0}J_{n+\frac{1}{2}}(k_i r_0)P_n^m(\cos\theta)\mathrm{e}^{jm\varphi} = a_{mn}^a \frac{\mu_a \omega m}{r\sin\theta}\sqrt{k_a r_0}H_{n+\frac{1}{2}}^{(2)}(k_a r_0)P_n^m(\cos\theta)\mathrm{e}^{jm\varphi}$$

$$a_{mn}^i \mu_i \sqrt{k_i}J_{n+\frac{1}{2}}(k_i r_0) = a_{mn}^a \mu_a \sqrt{k_a}H_{n+\frac{1}{2}}^{(2)}(k_a r_0)$$

$E_\varphi^i = E_\varphi^a$ gives

$$a_{mn}^i \frac{\mu_i j\omega}{r}\sqrt{k_i r_0}J_{n+\frac{1}{2}}(k_i r_0)\frac{\partial}{\partial\theta}P_n^m(\cos\theta)\mathrm{e}^{jm\varphi} = a_{mn}^a \frac{\mu_a j\omega}{r}\sqrt{k_a r_0}H_{n+\frac{1}{2}}^{(2)}(k_a r_0)\frac{\partial}{\partial\theta}P_n^m(\cos\theta)\mathrm{e}^{jm\varphi}$$

$$a_{mn}^i \mu_i \sqrt{k_i}J_{n+\frac{1}{2}}(k_i r_0) = a_{mn}^a \mu_a \sqrt{k_a}H_{n+\frac{1}{2}}^{(2)}(k_a r_0)$$

$H_\theta^i = H_\theta^a$ gives

$$a_{mn}^i \frac{1}{r_0}\frac{\partial}{\partial r}\left(\sqrt{k_i r}J_{n+\frac{1}{2}}(k_i r)\right)_{r=r_0}\frac{\partial}{\partial\theta}P_n^m(\cos\theta)\mathrm{e}^{jm\varphi} = a_{mn}^a \frac{1}{r_0}\frac{\partial}{\partial r}\left(\sqrt{k_a r}H_{n+\frac{1}{2}}^{(2)}(k_a r)\right)_{r=r_0}\frac{\partial}{\partial\theta}P_n^m(\cos\theta)\mathrm{e}^{jm\varphi}$$

$$a_{mn}^i \frac{\partial}{\partial r}\left(\sqrt{k_i r}J_{n+\frac{1}{2}}(k_i r)\right)_{r=r_0} = a_{mn}^a \frac{\partial}{\partial r}\left(\sqrt{k_a r}H_{n+\frac{1}{2}}^{(2)}(k_a r)\right)_{r=r_0}$$

$H_\varphi^i = H_\varphi^a$ gives

$$a_{mn}^i \frac{jm}{r_0\sin\theta}\frac{\partial}{\partial r}\left(\sqrt{k_i r}J_{n+\frac{1}{2}}(k_i r)\right)_{r=r_0}P_n^m(\cos\theta)\mathrm{e}^{jm\varphi} = a_{mn}^a \frac{jm}{r_0\sin\theta}\frac{\partial}{\partial r}\left(\sqrt{k_a r}H_{n+\frac{1}{2}}^{(2)}(k_a r)\right)_{r=r_0}P_n^m(\cos\theta)\mathrm{e}^{jm\varphi}$$

$$a_{mn}^i \frac{\partial}{\partial r}\left(\sqrt{k_i r}J_{n+\frac{1}{2}}(k_i r)\right)_{r=r_0} = a_{mn}^a \frac{\partial}{\partial r}\left(\sqrt{k_a r}H_{n+\frac{1}{2}}^{(2)}(k_a r)\right)_{r=r_0}$$

Rearranging

$$a_{mn}^i \mu_i \sqrt{k_i}J_{n+\frac{1}{2}}(k_i r_0) - a_{mn}^a \mu_a \sqrt{k_a}H_{n+\frac{1}{2}}^{(2)}(k_a r_0) = 0$$ and

$$a_{mn}^{i}\frac{\partial}{\partial r}\left(\sqrt{k_i r}J_{n+\frac{1}{2}}(k_i r)\right)_{r=r_0}-a_{mn}^{a}\frac{\partial}{\partial r}\left(\sqrt{k_a r}H_{n+\frac{1}{2}}^{(2)}(k_a r)\right)_{r=r_0}=0$$

$$\begin{pmatrix} \mu_i\sqrt{k_i}J_{n+\frac{1}{2}}(k_i r_0) & -\mu_a\sqrt{k_a}H_{n+\frac{1}{2}}^{(2)}(k_a r_0) \\ \frac{\partial}{\partial r}\left(\sqrt{k_i r}J_{n+\frac{1}{2}}(k_i r)\right)_{r=r_0} & -\frac{\partial}{\partial r}\left(\sqrt{k_a r}H_{n+\frac{1}{2}}^{(2)}(k_a r)\right)_{r=r_0}\end{pmatrix}\begin{pmatrix} a_{mn}^{i} \\ a_{mn}^{a}\end{pmatrix}=\begin{pmatrix}0\\0\end{pmatrix}$$

$$\mu_i\sqrt{k_i}J_{n+\frac{1}{2}}(k_i r_0)\frac{\partial}{\partial r}\left(\sqrt{k_a r}H_{n+\frac{1}{2}}^{(2)}(k_a r)\right)_{r=r_0}=\mu_a\sqrt{k_a}H_{n+\frac{1}{2}}^{(2)}(k_a r_0)\frac{\partial}{\partial r}\left(\sqrt{k_i r}J_{n+\frac{1}{2}}(k_i r)\right)_{r=r_0}$$

$$\frac{\frac{\partial}{\partial r}\left(\sqrt{k_a r}H_{n+\frac{1}{2}}^{(2)}(k_a r)\right)_{r=r_0}}{\frac{\partial}{\partial r}\left(\sqrt{k_i r}J_{n+\frac{1}{2}}(k_i r)\right)_{r=r_0}}=\frac{\mu_a\sqrt{k_a}H_{n+\frac{1}{2}}^{(2)}(k_a r_0)}{\mu_i\sqrt{k_i}J_{n+\frac{1}{2}}(k_i r_0)}$$

factor $\frac{k_a}{k_i}$ as a left-over on the LHS to be considered, thus in end effect only the ratio $\frac{k_i^{\frac{1}{2}}}{k_a^{\frac{1}{2}}}$ is involved (*improved reading* the ratio is of course dimensionless)

The spherical hollow cavity resonator

$\Pi_r(r,\theta,\varphi;t)=\sqrt{kr}J_{n+\frac{1}{2}}(kr)P_n^m(\cos\theta)\mathrm{e}^{\pm jm\varphi}$

$k^2=-j\omega\mu(\sigma+j\omega\varepsilon)$ assuming $\sigma\to\infty$

The electrical field vector is to be perpendicular to the spherical surface i.e. @ $r=r_0$, the radius of the cavity resonator, $E_\theta=0$ and $E_\varphi=0$

Boundary conditions @ $r=r_0$

$J(.)$ to ensure finite limits at zero.

TM $\frac{\partial}{\partial r}\left(\sqrt{kr}J_{n+\frac{1}{2}}(kr)\right)_{r=r_0}=0$

TE $J_{n+\frac{1}{2}}(kr)_{r=r_0}=0$

For TM, letting $\rho=kr$, $\partial\rho=k\partial r$ to write $\frac{\partial}{\partial r}\left(\sqrt{kr}J_{n+\frac{1}{2}}(kr)\right)=k\left(\frac{1}{2}\rho^{\frac{-1}{2}}J_{n+\frac{1}{2}}(\rho)+\sqrt{\rho}J'_{n+\frac{1}{2}}(\rho)\right)$ where the derivative is w.r.t. the argument ρ. Using the standard relationship $J'_\nu(\rho)=\frac{-\nu}{\rho}J_\nu(\rho)+J_{\nu-1}(\rho)$ leads to $\frac{\partial}{\partial r}\left(\sqrt{kr}J_{n+\frac{1}{2}}(kr)\right)=\frac{k}{\sqrt{\rho}}\left(-nJ_{n+\frac{1}{2}}(\rho)+\rho J_{n-\frac{1}{2}}(\rho)\right)$, which on equating to zero yields the transcendental equation $\frac{J_{n+\frac{1}{2}}(\rho)}{J_{n-\frac{1}{2}}(\rho)}=\frac{\rho}{n}$ that is to be solved for the roots ρ. Denote the $\nu-th$ root by $b_{n+\frac{1}{2},\nu}$ then $\lambda=\frac{2\pi}{k}=\frac{2\pi}{\frac{\rho}{r}}=\frac{2\pi r}{b_{n+\frac{1}{2},\nu}}|_{r=r_0}$. The cavity resonator is capable of sustaining oscillation at wavelengths $\lambda=\frac{2\pi r_0}{b_{n+\frac{1}{2},\nu}}$ in $\mathrm{TM}_{mn\nu}$ modes.

Likewise, for TE denote the roots of $J_{n+\frac{1}{2}}(\rho)=0$ by $c_{n+\frac{1}{2},\nu}$. $\text{TE}_{mn\nu}$ modes are sustainable at wavelengths $\lambda=\dfrac{2\pi r_0}{c_{n+\frac{1}{2},\nu}}$.

The spherical radiator

Boundary conditions @ $r=r_0$

$H^{(2)}(.)$ to ensure zero at ∞.

TM $\dfrac{\partial}{\partial r}\left(\sqrt{kr}\,H^{(2)}_{n+\frac{1}{2}}(kr)\right)_{r=r_0}=0$

TE $H^{(2)}_{n+\frac{1}{2}}(kr)_{r=r_0}=0$

Taking the lowest index $n=1$ as example, the Hankel function is assembled from the corresponding Bessel and Neumann functions.

Using the known relationships $\frac{2n}{x}Z_n=Z_{n-1}+Z_{n+1}$,
and $J_{\frac{1}{2}}(x)=\sqrt{\frac{2}{\pi x}}\sin x$, and $J_{\frac{-1}{2}}(x)=\sqrt{\frac{2}{\pi x}}\cos x$ to arrive at

$$J_{\frac{3}{2}}(x)=\sqrt{\tfrac{2}{\pi x}}\left(\tfrac{\sin x}{x}-\cos x\right)$$

$$J_{\frac{-3}{2}}(x)=-\sqrt{\tfrac{2}{\pi x}}\left(\tfrac{\cos x}{x}+\sin x\right)$$

Also

$$N_\nu(x)=\frac{\cos\nu\pi\ J_\nu(x)-J_{-\nu}(x)}{\sin\nu\pi}$$

$$N_{\frac{3}{2}}(x)=\frac{\cos\frac{3}{2}\pi\ J_{\frac{3}{2}}(x)-J_{\frac{-3}{2}}(x)}{\sin\frac{3}{2}\pi}=\frac{0-J_{\frac{-3}{2}}(x)}{-1}$$

$$N_{\frac{3}{2}}(x)=J_{\frac{-3}{2}}(x)$$

$$N_{\frac{3}{2}}(x)=-\sqrt{\tfrac{2}{\pi x}}\left(\tfrac{\cos x}{x}+\sin x\right)$$

Definition $H^{(2)}_{\frac{3}{2}}(x)=J_{\frac{3}{2}}(x)-j\,N_{\frac{3}{2}}(x)$

Proceeding

$$H^{(2)}_{\frac{3}{2}}(x)=\sqrt{\tfrac{2}{\pi x}}\left(\tfrac{\sin x}{x}-\cos x\right)+j\sqrt{\tfrac{2}{\pi x}}\left(\tfrac{\cos x}{x}+\sin x\right)$$

$$=\sqrt{\tfrac{2}{\pi x}}\left((\tfrac{j\cos x}{x}-\cos x)+(\tfrac{\sin x}{x}+j\sin x)\right)$$

$$=\sqrt{\tfrac{2}{\pi x}}\left((\tfrac{j\cos x}{x}-\cos x)-(\tfrac{-\sin x}{x}-j\sin x)\right)$$

$$=\sqrt{\tfrac{2}{\pi x}}\left((\tfrac{j\cos x}{x}-\cos x)-(\tfrac{j\,j\sin x}{x}-j\sin x)\right)$$

$$=\sqrt{\tfrac{2}{\pi x}}\left(\cos x(\tfrac{j}{x}-1)-j\sin x(\tfrac{j}{x}-1)\right)$$

$$=\sqrt{\tfrac{2}{\pi x}}\left((\cos x-j\sin x)(\tfrac{j}{x}-1)\right)$$

$$=\sqrt{\tfrac{2}{\pi x}}\,\mathrm{e}^{-jx}(\tfrac{j}{x}-1)$$

For TM, again letting $\rho = kr$ to write $k\frac{\partial}{\partial\rho}\left(\sqrt{\rho}H_{\frac{3}{2}}^{(2)}(\rho)\right) = k\frac{\partial}{\partial\rho}\left(\sqrt{\frac{2}{\pi}}\,\mathrm{e}^{-j\rho}(\frac{j}{\rho}-1)\right)$ as the derivative is w.r.t. the argument.

The boundary condition becomes

$$\frac{\partial}{\partial\rho}\mathrm{e}^{-j\rho}(\tfrac{j}{\rho}-1)\overset{\downarrow}{=}0$$

$$-j\,\mathrm{e}^{-j\rho}(\tfrac{j}{\rho}-1)+\mathrm{e}^{-j\rho}(\tfrac{-j}{\rho^2})=0$$

$$-j(\tfrac{j}{\rho}-1)+(\tfrac{-j}{\rho^2})=0$$

$$\tfrac{1}{\rho}+j-\tfrac{j}{\rho^2}=0$$

$$\rho+j\rho^2-j=0$$

$$\rho^2-j\rho-1=0$$

The roots work out to be

$$\rho_{11},\rho_{12}=\frac{-(-j)\pm\sqrt{3}}{2}$$

From $\rho = kr_0 = \sqrt{\varepsilon\mu}\;\omega\, r_0$

$$\omega_{1*}=\frac{\rho_{1*}}{\sqrt{\varepsilon\mu}}\frac{1}{r_0}=\frac{\pm\frac{\sqrt{3}}{2}+j\,0.5}{\sqrt{\varepsilon\mu}}\frac{1}{r_0}$$

The time dependency may be written as

$$\mathrm{e}^{j\omega t}=\mathrm{e}^{\frac{-0.5}{\sqrt{\varepsilon\mu}}\frac{1}{r_0}t\pm j\frac{0.866}{\sqrt{\varepsilon\mu}}\frac{1}{r_0}t}$$

The radiation is not free of damping. The wavelength may be given as

$$\lambda=\frac{2\pi}{k}=\frac{2\pi}{\rho/r_0}=\frac{2\pi}{\frac{\sqrt{3}}{2}}r_0\approx 7.255r_0$$

For the TE mode under consideration, the radiator is exhibiting a static damping behaviour. In particular, the root for the Hankel function is found at $\rho = 0 + j$, thus

$$\mathrm{e}^{j\omega t}=\mathrm{e}^{\frac{-1}{\sqrt{\varepsilon\mu}}\frac{1}{r_0}t}$$

The spherical antenna (TM waves)

Assuming $\frac{\partial}{\partial\varphi}\equiv 0$ i.e. $m=0$. Proceeding for TM

$$E_r = a_{0n}k^2\frac{n(n+1)}{(kr)^{\frac{3}{2}}}H_{n+\frac{1}{2}}^{(2)}(kr)P_n(\cos\theta)$$

$$E_\theta = a_{0n}\frac{1}{r}\frac{\partial}{\partial r}\left(\sqrt{kr}H_{n+\frac{1}{2}}^{(2)}(kr)\right)\frac{\partial}{\partial\theta}P_n(\cos\theta) \qquad \text{generic expression}$$

$$E_\varphi = 0$$

$$H_r = 0$$

$$H_\theta = 0$$

$$H_\varphi = a_{0n}\frac{-\varepsilon j\omega}{r}\sqrt{kr}H_{n+\frac{1}{2}}^{(2)}(kr)\frac{\partial}{\partial\theta}P_n(\cos\theta)$$

Recalling $k^2=\varepsilon\mu\omega^2$

Using $Z_n' = -\frac{n}{x}Z_n + Z_{n-1}$ to have

$$\frac{\partial}{\partial r}\left(\sqrt{kr}\,H^{(2)}_{n+\frac{1}{2}}(kr)\right)=\frac{k}{\sqrt{kr}}\left(-nH^{(2)}_{n+\frac{1}{2}}(kr)+krH^{(2)}_{n-\frac{1}{2}}(kr)\right)$$

Also recalling

$$\frac{\partial}{\partial\theta}P_n(\cos\theta)=-P_n^1(\cos\theta)$$

to arrive at the following

$$E_r=a_{0n}k^2\frac{n(n+1)}{(kr)^{\frac{3}{2}}}H^{(2)}_{n+\frac{1}{2}}(kr)P_n(\cos\theta)$$

$$E_\theta=a_{0n}\frac{1}{r}\frac{k}{\sqrt{kr}}\left(nH^{(2)}_{n+\frac{1}{2}}(kr)-krH^{(2)}_{n-\frac{1}{2}}(kr)\right)P_n^1(\cos\theta)$$

$$H_\varphi=a_{0n}\frac{\varepsilon j\omega}{r}\sqrt{kr}\,H^{(2)}_{n+\frac{1}{2}}(kr)P_n^1(\cos\theta)$$

Consider $f(\theta)$ the dependency on θ of E_θ at the boundary surface $r=r_0$. It can be expanded into a series $f(\theta)=\sum_{n=1}^{\infty}b_nP_n^1(\cos\theta)$ where the coefficients may be given as $b_n=\frac{2n+1}{2n(n+1)}\int_0^\pi f(\theta)P_n^1(\cos\theta)\sin\theta\,d\theta=\frac{2n+1}{2n(n+1)}\int_0^\pi f(\theta)P_n^1(\cos\theta)\frac{1}{r_0}r_0\sin\theta\,d\theta$. As neither $\sin\theta$ nor $P_n^1(\cos\theta)$ would, respectively, deviate from 1 and $P_n^1(0)$ over the integration interval of interest, they may be taken out of the integral $b_n\approx\frac{2n+1}{2n(n+1)r_0}P_n^1(0)\int_{-r_0\alpha}^{+r_0\alpha}f(\theta)\,dl=\frac{2n+1}{2n(n+1)r_0}P_n^1(0)U_0$

On the other hand superposition over the indices yields

$$E_\theta=\sum_{n=1}^{\infty}a_{0n}\frac{1}{r}\frac{k}{\sqrt{kr}}\left(nH^{(2)}_{n+\frac{1}{2}}(kr)-krH^{(2)}_{n-\frac{1}{2}}(kr)\right)P_n^1(\cos\theta)\text{ at } r=r_0$$

$E_\theta=\sum_{n=1}^{\infty}a_{0n}\frac{1}{r_0}\frac{k}{\sqrt{kr_0}}\left(nH^{(2)}_{n+\frac{1}{2}}(kr_0)-kr_0H^{(2)}_{n-\frac{1}{2}}(kr_0)\right)P_n^1(\cos\theta)$ Comparing terms for the two series gives the coefficients a_{0n}

$a_{0n}\frac{1}{r_0}\frac{k}{\sqrt{kr_0}}\left(nH^{(2)}_{n+\frac{1}{2}}(kr_0)-kr_0H^{(2)}_{n-\frac{1}{2}}(kr_0)\right)=\frac{2n+1}{2n(n+1)r_0}P_n^1(0)U_0$ With the availability of the a_{0n}, the field is now known. (Calculations may proceed to obtain any params of interest.)

Interesting to consider $n=1$ where $H^{(2)}_{\frac{3}{2}}(kr)=\sqrt{\frac{2}{\pi kr}}\,e^{-jkr}(\frac{j}{kr}-1)$ as shown above.

Derivative $\frac{\partial}{\partial r}\left(\sqrt{kr}\,H^{(2)}_{\frac{3}{2}}(kr)\right)$ where $H^{(2)}_{\frac{3}{2}}(kr)=\sqrt{\frac{2}{\pi kr}}\,e^{-jkr}(\frac{j}{kr}-1)$, from which $\sqrt{kr}\,H^{(2)}_{\frac{3}{2}}(kr)=\sqrt{\frac{2}{\pi}}\,e^{-jkr}(\frac{j}{kr}-1)$, thus $\frac{\partial}{\partial r}\left(\sqrt{kr}\,H^{(2)}_{\frac{3}{2}}(kr)\right)=\frac{\partial}{\partial r}\left(\sqrt{\frac{2}{\pi}}\,e^{-jkr}(\frac{j}{kr}-1)\right)$. Letting $\rho=kr$, $\partial\rho=k\partial r$

$$\frac{\partial}{\partial r}\left(\sqrt{\frac{2}{\pi}}\,e^{-jkr}(\frac{j}{kr}-1)\right)=k\frac{\partial}{\partial\rho}\left(\sqrt{\frac{2}{\pi}}\,e^{-j\rho}(\frac{j}{\rho}-1)\right)$$

$$=k\sqrt{\frac{2}{\pi}}\left(-je^{-j\rho}(\frac{j}{\rho}-1)+e^{-j\rho}(\frac{-j}{\rho^2})\right)$$

$$= k\sqrt{\tfrac{2}{\pi}}\,\mathrm{e}^{-j\rho}\left(-j(\tfrac{j}{\rho}-1)+(\tfrac{-j}{\rho^2})\right)$$
$$= k\sqrt{\tfrac{2}{\pi}}\,\mathrm{e}^{-j\rho}\left(\tfrac{-j^2}{\rho}+j-j\tfrac{1}{\rho^2}\right)$$
$$= k\sqrt{\tfrac{2}{\pi}}\,\mathrm{e}^{-j\rho}\left(j+\tfrac{1}{\rho}-j\tfrac{1}{\rho^2}\right)$$
$$= k\sqrt{\tfrac{2}{\pi}}\,\mathrm{e}^{-jkr}\left(j+\tfrac{1}{kr}-j\tfrac{1}{(kr)^2}\right)$$

Generic

$$E_r = a_{01}k^2\frac{2}{(kr)^{\frac{3}{2}}}H^{(2)}_{\frac{3}{2}}(kr)\cos\theta$$
$$E_\theta = -a_{01}\frac{1}{r}\frac{\partial}{\partial r}\left(\sqrt{kr}\,H^{(2)}_{\frac{3}{2}}(kr)\right)\sin\theta$$

grasping back to the generic expression above

$$H_\varphi = a_{01}\frac{\varepsilon j\omega}{r}\sqrt{kr}\,H^{(2)}_{\frac{3}{2}}(kr)\sin\theta$$

Take 1

$$E_r = a_{01}k^2\frac{2}{(kr)^{\frac{3}{2}}}\sqrt{\tfrac{2}{\pi kr}}\,\mathrm{e}^{-jkr}(\tfrac{j}{kr}-1)\cos\theta$$
$$E_\theta = -a_{01}\frac{1}{r}k\sqrt{\tfrac{2}{\pi}}\,\mathrm{e}^{-jkr}\left(j+\tfrac{1}{kr}-j\tfrac{1}{(kr)^2}\right)\sin\theta$$
$$H_\varphi = a_{01}\frac{\varepsilon j\omega}{r}\sqrt{kr}\sqrt{\tfrac{2}{\pi kr}}\,\mathrm{e}^{-jkr}(\tfrac{j}{kr}-1)\sin\theta$$

Take 2

$$E_r = a_{01}k^2\frac{2}{(kr)^2}\sqrt{\tfrac{2}{\pi}}\,\mathrm{e}^{-jkr}(\tfrac{j}{kr}-1)\cos\theta$$
$$E_\theta = -a_{01}\frac{1}{kr}k^2\sqrt{\tfrac{2}{\pi}}\,\mathrm{e}^{-jkr}\left(j+\tfrac{1}{kr}-j\tfrac{1}{(kr)^2}\right)\sin\theta$$
$$H_\varphi = a_{01}\frac{\varepsilon j\omega}{kr}k\sqrt{\tfrac{2}{\pi}}\,\mathrm{e}^{-jkr}(\tfrac{j}{kr}-1)\sin\theta$$

Take 3

$$E_r = a_{01}k^2 2\sqrt{\tfrac{2}{\pi}}\,\mathrm{e}^{-jkr}(\tfrac{-1}{(kr)^2}+\tfrac{j}{(kr)^3})\cos\theta$$
$$E_\theta = -a_{01}k^2\sqrt{\tfrac{2}{\pi}}\,\mathrm{e}^{-jkr}\left(\tfrac{j}{kr}+\tfrac{1}{(kr)^2}-\tfrac{j}{(kr)^3}\right)\sin\theta$$
$$H_\varphi = a_{01}\varepsilon j\omega k\sqrt{\tfrac{2}{\pi}}\,\mathrm{e}^{-jkr}(-\tfrac{1}{kr}+\tfrac{j}{(kr)^2})\sin\theta$$

Take 4

$$E_r = a_{01}jk^2 2\sqrt{\tfrac{2}{\pi}}\,\mathrm{e}^{-jkr}(\tfrac{j}{(kr)^2}+\tfrac{1}{(kr)^3})\cos\theta$$
$$E_\theta = a_{01}jk^2\sqrt{\tfrac{2}{\pi}}\,\mathrm{e}^{-jkr}\left(-\tfrac{1}{kr}+\tfrac{j}{(kr)^2}+\tfrac{1}{(kr)^3}\right)\sin\theta$$
$$H_\varphi = a_{01}\varepsilon j\omega k\sqrt{\tfrac{2}{\pi}}\,\mathrm{e}^{-jkr}(-\tfrac{1}{kr}+\tfrac{j}{(kr)^2})\sin\theta$$

Except for the constants, the field is identical to that of an elementary dipole antenna. Both cases show r raised to the same powers in their corresponding terms of the denominators.

§§§

Improved reading, a reminder, not a correction

Klystrons and magnetrons work on the same principle that applies for hollow cavity resonators, i.e. they work at some dominant eigenfrequencies. They differ in the way the electromagnetic field is excited and channeled through a device, in the way efficiency criteria are to be attained, how ease of operation is achieved (e.g. in regard to frequency stability, trimming, mechanical sizes and maintenance, etc.) and, of course, in the applications they are designed for. [Jordan, "*Reference Data for Engineers*"]

Checking units or physical dimensions for consistency for the equation $grad\ \phi - curl\ curl\ \mathbf{\Pi} + k^2\mathbf{\Pi} = 0$

Improved reading, physical units or dimensions for the Hertz vectors. Before proceeding, recall

$$\omega = 2\pi f = 2\pi \frac{c}{\lambda} \qquad \lambda = 2\pi \frac{c}{\omega} = 2\pi \frac{1}{\frac{\omega}{c}} = 2\pi \frac{1}{k} \qquad \text{where } k = \frac{\omega}{c} = \omega\sqrt{\varepsilon\mu} \quad \text{there } [k] = \frac{\frac{1}{s}}{\frac{m}{s}} = \frac{1}{m}$$

operator $[.]$ takes the dimension of operand in the following, as usual.

TM: Consider $\mathbf{H} = (\sigma + j\omega\varepsilon) curl\mathbf{\Pi}_e$ or just $\mathbf{H} = j\omega\varepsilon\, curl\mathbf{\Pi}_e$, $[\mathbf{H}] = [\omega\varepsilon][curl\mathbf{\Pi}_e]$, $[\mathbf{H}] = \frac{1}{s}[\varepsilon]\frac{1}{m}[\mathbf{\Pi}_e]$, $[\mathbf{H}] = \frac{1}{s}\frac{As}{Vm}\frac{1}{m}[\mathbf{\Pi}_e]$, $[\mathbf{H}] = \frac{A}{Vm^2}[\mathbf{\Pi}_e]$. With $[\mathbf{H}] = \frac{A}{m}$ write $\frac{A}{m} = \frac{A}{Vm^2}[\mathbf{\Pi}_e]$ to get $[\mathbf{\Pi}_e] = Vm$

TE: Consider $\mathbf{E} = -j\omega\mu\, curl\mathbf{\Pi}_m$, $[\mathbf{E}] = [\omega\mu][curl\mathbf{\Pi}_m]$, $[\mathbf{E}] = \frac{1}{s}[\mu]\frac{1}{m}[\mathbf{\Pi}_m]$, $[\mathbf{E}] = \frac{1}{s}\frac{Vs}{Am}\frac{1}{m}[\mathbf{\Pi}_m]$, $[\mathbf{E}] = \frac{V}{Am^2}[\mathbf{\Pi}_m]$. With $[\mathbf{E}] = \frac{V}{m}$ write $\frac{V}{m} = \frac{V}{Am^2}[\mathbf{\Pi}_m]$ to get $[\mathbf{\Pi}_m] = Am$

Cylindrical Coordinates

Procedure

The Maxwell Eqs. led to an intractable equation for the general case namely
$grad\ \phi - curl\ curl\ \mathbf{\Pi} + k^2\mathbf{\Pi} = 0$
where $\mathbf{\Pi}$ is the Hertz vector.

If $\mathbf{\Pi}$ were known, the field vectors would be computed from
$\mathbf{H} = (\sigma + j\varepsilon\omega)\ curl\ \mathbf{\Pi}$
$\mathbf{E} = k^2\ \mathbf{\Pi} + grad\ \phi$
or
$\mathbf{E} = -j\mu\omega\ curl\ \mathbf{\Pi}$
$\mathbf{H} = k^2\ \mathbf{\Pi} + grad\ \phi$

Simplifying Ansätze have been made. Usually, $\mathbf{\Pi}$ is assumed to have only one distinct non-zero component. Also in accommodating some symmetric geometry, such as that of a concentric cylinder or a sphere, the choice for ϕ is to alleviate the efforts as far as possible. The Ansätze led to well known differential equations of Physics, with known solutions for $\mathbf{\Pi}$. (This text is showing one or two examples for the latter mentioned.)

A list is compiled for the expressions for the fields, taking $Z_m(sr)$ as example in the following
$\Pi_z(z,r,\varphi;t) = Z_m(\sqrt{k^2-\beta^2}\,r)\mathrm{e}^{\pm jm\varphi}\mathrm{e}^{\pm j\beta z}\mathrm{e}^{j\omega t}$
$k^2 = -j\omega\mu(\sigma + j\omega\varepsilon)$ involving frequency and material constants, and $s^2 = k^2 - \beta^2$

TM

$$E_z = (k^2-\beta^2)\Pi_z \qquad H_z = 0$$

$$E_r = -j\beta\frac{\partial \Pi_z}{\partial r} \qquad H_r = \frac{jk^2}{\omega\mu r}\frac{\partial \Pi_z}{\partial \varphi}$$

$$E_\varphi = -j\frac{\beta}{r}\frac{\partial \Pi_z}{\partial \varphi} \qquad H_\varphi = -\frac{jk^2}{\omega\mu}\frac{\partial \Pi_z}{\partial r}$$

TE

$$E_z = 0 \qquad H_z = (k^2-\beta^2)\Pi_z$$

$$E_r = -\frac{j\omega\mu}{r}\frac{\partial \Pi_z}{\partial \varphi} \qquad H_r = -j\beta\frac{\partial \Pi_z}{\partial r}$$

$$E_\varphi = j\omega\mu\frac{\partial \Pi_z}{\partial r} \qquad H_\varphi = -j\frac{\beta}{r}\frac{\partial \Pi_z}{\partial \varphi}$$

Proceeding for TM from using $\Pi_z(z,r,\varphi;t) = J_n(sr)\mathrm{e}^{-jn\varphi}\mathrm{e}^{j(\omega t-\beta z)}$

$$E_z = (k^2-\beta^2)\Pi_z \qquad E_z = s^2 J_n(sr)\mathrm{e}^{-jn\varphi}\mathrm{e}^{j(\omega t-\beta z)}$$

$$E_r = -j\beta\frac{\partial \Pi_z}{\partial r} \qquad E_r = -j\beta\frac{\partial J_n(sr)\mathrm{e}^{-jn\varphi}\mathrm{e}^{j(\omega t-\beta z)}}{\partial r} = -j\beta s J_n'(sr)\mathrm{e}^{-jn\varphi}\mathrm{e}^{j(\omega t-\beta z)}$$

$$E_\varphi = -j\frac{\beta}{r}\frac{\partial \Pi_z}{\partial \varphi} \qquad E_\varphi = -j\frac{\beta}{r}\frac{\partial J_n(sr)\mathrm{e}^{-jn\varphi}\mathrm{e}^{j(\omega t-\beta z)}}{\partial \varphi} = -j\frac{\beta}{r}J_n(sr)(-jn)\mathrm{e}^{-jn\varphi}\mathrm{e}^{j(\omega t-\beta z)}$$

$$H_z = 0$$

$$H_r = \frac{jk^2}{\omega\mu r}\frac{\partial \Pi_z}{\partial \varphi} \qquad H_r = \frac{jk^2}{\omega\mu r}\frac{\partial J_n(sr)\mathrm{e}^{-jn\varphi}\mathrm{e}^{j(\omega t-\beta z)}}{\partial \varphi} = \frac{jk^2}{\omega\mu r}J_n(sr)(-jn)\mathrm{e}^{-jn\varphi}\mathrm{e}^{j(\omega t-\beta z)}$$

$$H_\varphi = -\frac{jk^2}{\omega\mu}\frac{\partial \Pi_z}{\partial r} \qquad H_\varphi = -\frac{jk^2}{\omega\mu}\frac{\partial J_n(sr)\mathrm{e}^{-jn\varphi}\mathrm{e}^{j(\omega t-\beta z)}}{\partial r} = -\frac{jk^2}{\omega\mu}sJ_n'(sr)\mathrm{e}^{-jn\varphi}\mathrm{e}^{j(\omega t-\beta z)}$$

Proceeding for TE from using $\Pi_z(z,r,\varphi;t) = J_n(sr)\mathrm{e}^{-jn\varphi}\mathrm{e}^{j(\omega t-\beta z)}$

$$E_z = 0$$

$$E_r = -\frac{j\omega\mu}{r}\frac{\partial \Pi_z}{\partial \varphi} \qquad E_r = -\frac{j\omega\mu}{r}\frac{\partial J_n(sr)\mathrm{e}^{-jn\varphi}\mathrm{e}^{j(\omega t-\beta z)}}{\partial \varphi} = -\frac{j\omega\mu}{r}J_n(sr)(-jn)\mathrm{e}^{-jn\varphi}\mathrm{e}^{j(\omega t-\beta z)}$$

$$E_\varphi = j\omega\mu\frac{\partial \Pi_z}{\partial r} \qquad E_\varphi = j\omega\mu\frac{\partial J_n(sr)\mathrm{e}^{-jn\varphi}\mathrm{e}^{j(\omega t-\beta z)}}{\partial r} = j\omega\mu s J_n'(sr)\mathrm{e}^{-jn\varphi}\mathrm{e}^{j(\omega t-\beta z)}$$

$$H_z = (k^2-\beta^2)\Pi_z \qquad H_z = s^2 J_n(sr)\mathrm{e}^{-jn\varphi}\mathrm{e}^{j(\omega t-\beta z)}$$

$$H_r = -j\beta\frac{\partial \Pi_z}{\partial r} \qquad H_r = -j\beta\frac{\partial J_n(sr)\mathrm{e}^{-jn\varphi}\mathrm{e}^{j(\omega t-\beta z)}}{\partial r} = -j\beta s J_n'(sr)\mathrm{e}^{-jn\varphi}\mathrm{e}^{j(\omega t-\beta z)}$$

$$H_\varphi = -j\frac{\beta}{r}\frac{\partial \Pi_z}{\partial \varphi} \qquad H_\varphi = -j\frac{\beta}{r}\frac{\partial J_n(sr)\mathrm{e}^{-jn\varphi}\mathrm{e}^{j(\omega t-\beta z)}}{\partial \varphi} = -j\frac{\beta}{r}J_n(sr)(-jn)\mathrm{e}^{-jn\varphi}\mathrm{e}^{j(\omega t-\beta z)}$$

Boundary conditions

Consider a long concentric cylinder with material constants $\varepsilon_i\ \mu_i\ \sigma_i$ in medium characterized by $\varepsilon_a\ \mu_a\ \sigma_a$.

It is asked what waveforms can propagate *along* the cylinder. The superposition of TM (coefficients $a_n^{i,a}$) and TE (coefficients $b_n^{i,a}$) would be a sensible Ansatz for a general solution

For the field within the cylinder (index i, the derivative $J_n'(s_i r)$ is wrt the argument)

$$E_z^i = \sum_n a_n^i s_i^2 J_n(s_i r)\mathrm{e}^{-jn\varphi}\mathrm{e}^{j(\omega t-\beta z)} + 0_{TE}$$

$$E_r^i = \sum_n \left(-a_n^i j\beta s_i J_n'(s_i r) - b_n^i \frac{j\omega\mu_i}{r}J_n(s_i r)(-jn)\right)\mathrm{e}^{-jn\varphi}\mathrm{e}^{j(\omega t-\beta z)}$$

$$E_\varphi^i = \sum_n \left(-a_n^i j\frac{\beta}{r}J_n(s_i r)(-jn) + b_n^i j\omega\mu_i s_i J_n'(s_i r)\right)\mathrm{e}^{-jn\varphi}\mathrm{e}^{j(\omega t-\beta z)}$$

$$H_z^i = 0_{TM} + \sum_n b_n^i s_i^2 J_n(s_i r)\mathrm{e}^{-jn\varphi}\mathrm{e}^{j(\omega t-\beta z)}$$

$$H_r^i = \sum_n \left(a_n^i \frac{jk_i^2}{\omega\mu_i r}J_n(s_i r)(-jn) - b_n^i j\beta s_i J_n'(s_i r)\right)\mathrm{e}^{-jn\varphi}\mathrm{e}^{j(\omega t-\beta z)}$$

$$H_\varphi^i = \sum_n \left(-a_n^i \frac{jk_i^2}{\omega\mu_i} s_i J_n'(s_i r) - b_n^i j \frac{\beta}{r} J_n(s_i r)(-jn) \right) \mathrm{e}^{-jn\varphi} \mathrm{e}^{j(\omega t - \beta z)}$$

Cleaning up

$$E_z^i = \sum_n a_n^i s_i^2 J_n(s_i r) \mathrm{e}^{-jn\varphi} \mathrm{e}^{j(\omega t - \beta z)} + 0_{TE}$$

$$E_r^i = \sum_n \left(-a_n^i j\beta s_i J_n'(s_i r) - b_n^i \frac{\omega\mu_i}{r} n J_n(s_i r) \right) \mathrm{e}^{-jn\varphi} \mathrm{e}^{j(\omega t - \beta z)}$$

$$E_\varphi^i = \sum_n \left(-a_n^i \frac{\beta}{r} n J_n(s_i r) + b_n^i j\omega\mu_i s_i J_n'(s_i r) \right) \mathrm{e}^{-jn\varphi} \mathrm{e}^{j(\omega t - \beta z)}$$

$$H_z^i = 0_{TM} + \sum_n b_n^i s_i^2 J_n(s_i r) \mathrm{e}^{-jn\varphi} \mathrm{e}^{j(\omega t - \beta z)}$$

$$H_r^i = \sum_n \left(a_n^i \frac{k_i^2}{\omega\mu_i r} n J_n(s_i r) - b_n^i j\beta s_i J_n'(s_i r) \right) \mathrm{e}^{-jn\varphi} \mathrm{e}^{j(\omega t - \beta z)}$$

$$H_\varphi^i = \sum_n \left(-a_n^i \frac{jk_i^2}{\omega\mu_i} s_i J_n'(s_i r) - b_n^i \frac{\beta}{r} n J_n(s_i r) \right) \mathrm{e}^{-jn\varphi} \mathrm{e}^{j(\omega t - \beta z)}$$

The field surrounding the cylinder (index a, the derivative $H_n'^{(2)}(s_a r)$ is wrt the argument) reads

$$E_z^a = \sum_n a_n^a s_a^2 H_n^{(2)}(s_a r) \mathrm{e}^{-jn\varphi} \mathrm{e}^{j(\omega t - \beta z)} + 0_{TE}$$

$$E_r^a = \sum_n \left(-a_n^a j\beta s_a H_n'^{(2)}(s_a r) - b_n^a \frac{\omega\mu_a}{r} n H_n^{(2)}(s_a r) \right) \mathrm{e}^{-jn\varphi} \mathrm{e}^{j(\omega t - \beta z)}$$

$$E_\varphi^a = \sum_n \left(-a_n^a \frac{\beta}{r} n H_n^{(2)}(s_a r) + b_n^a j\omega\mu_a s_a H_n'^{(2)}(s_a r) \right) \mathrm{e}^{-jn\varphi} \mathrm{e}^{j(\omega t - \beta z)}$$

$$H_z^a = 0_{TM} + \sum_n b_n^a s_a^2 H_n^{(2)}(s_a r) \mathrm{e}^{-jn\varphi} \mathrm{e}^{j(\omega t - \beta z)}$$

$$H_r^a = \sum_n \left(a_n^a \frac{k_a^2}{\omega\mu_a r} n H_n^{(2)}(s_a r) - b_n^a j\beta s_a H_n'^{(2)}(s_a r) \right) \mathrm{e}^{-jn\varphi} \mathrm{e}^{j(\omega t - \beta z)}$$

$$H_\varphi^a = \sum_n \left(-a_n^a \frac{jk_a^2}{\omega\mu_a} s_a H_n'^{(2)}(s_a r) - b_n^a \frac{\beta}{r} n H_n^{(2)}(s_a r) \right) \mathrm{e}^{-jn\varphi} \mathrm{e}^{j(\omega t - \beta z)}$$

The tangential components are to be continuous at $r = r_0$

$E_\varphi^i = E_\varphi^a$ translating into

$$-a_n^i \frac{\beta}{r_0} n J_n(s_i r_0) + b_n^i j\omega\mu_i s_i J_n'(s_i r_0) = -a_n^a \frac{\beta}{r_0} n H_n^{(2)}(s_a r_0) + b_n^a j\omega\mu_a s_a H_n'^{(2)}(s_a r_0)$$

$E_z^i = E_z^a$ translating into

$$a_n^i s_i^2 J_n(s_i r_0) = a_n^a s_a^2 H_n^{(2)}(s_a r_0)$$

$H_\varphi^i = H_\varphi^a$ translating into

$$-a_n^i \frac{jk_i^2}{\omega\mu_i} s_i J_n'(s_i r_0) - b_n^i \frac{\beta}{r_0} n J_n(s_i r_0) = -a_n^a \frac{jk_a^2}{\omega\mu_a} s_a H_n'^{(2)}(s_a r_0) - b_n^a \frac{\beta}{r_0} n H_n^{(2)}(s_a r_0)$$

$H_z^i = H_z^a$ translating into

$$b_n^i s_i^2 J_n(s_i r_0) = b_n^a s_a^2 H_n^{(2)}(s_a r_0)$$

Reshuffling

$$-a_n^i\frac{\beta}{r_0}nJ_n(s_ir_0)+b_n^i j\omega\mu_i s_iJ_n'(s_ir_0)+a_n^a\frac{\beta}{r_0}nH_n^{(2)}(s_ar_0)-b_n^a j\omega\mu_a s_aH_n'^{(2)}(s_ar_0)=0$$

$$a_n^i s_i^2J_n(s_ir_0)-a_n^a s_a^2H_n^{(2)}(s_ar_0)=0$$

$$-a_n^i\frac{jk_i^2}{\omega\mu_i}s_iJ_n'(s_ir_0)-b_n^i\frac{\beta}{r_0}nJ_n(s_ir_0)+a_n^a\frac{jk_a^2}{\omega\mu_a}s_aH_n'^{(2)}(s_ar_0)+b_n^a\frac{\beta}{r_0}nH_n^{(2)}(s_ar_0)=0$$

$$b_n^i s_i^2J_n(s_ir_0)-b_n^a s_a^2H_n^{(2)}(s_ar_0)=0$$

$$\begin{pmatrix} -\frac{\beta}{r_0}nJ_n(s_ir_0) & j\omega\mu_i s_iJ_n'(s_ir_0) & \frac{\beta}{r_0}nH_n^{(2)}(s_ar_0) & -j\omega\mu_a s_aH_n'^{(2)}(s_ar_0) \\ s_i^2J_n(s_ir_0) & 0 & -s_a^2H_n^{(2)}(s_ar_0) & 0 \\ -\frac{jk_i^2}{\omega\mu_i}s_iJ_n'(s_ir_0) & -b_n^i\frac{\beta}{r_0}nJ_n(s_ir_0) & \frac{jk_a^2}{\omega\mu_a}s_aH_n'^{(2)}(s_ar_0) & \frac{\beta}{r_0}nH_n^{(2)}(s_ar_0) \\ 0 & s_i^2J_n(s_ir_0) & 0 & -s_a^2H_n^{(2)}(s_ar_0) \end{pmatrix}\begin{pmatrix} a_n^i \\ b_n^i \\ a_n^a \\ b_n^a \end{pmatrix}=\begin{pmatrix} 0 \\ 0 \\ 0 \\ 0 \end{pmatrix}$$

Zero determinant is required for non-trivial solutions to be possible.
The complicated transcendental equation resulting out of this requirement would serve to determine β from $s_i=\sqrt{k_i^2-\beta^2}$ or from $s_a=\sqrt{k_a^2-\beta^2}$ where as a reminder recalling that $k_i^2=-j\omega\mu_i(\sigma_i+j\omega\varepsilon_i)$ and $k_a^2=-j\omega\mu_a(\sigma_a+j\omega\varepsilon_a)$, the transcendental equation representing one of three equations for the 3 unknown β s_i and s_a. When β has become available, the equation system would render itself solveable for the coefficients.

The cylindrical hollow cavity resonator

The phenomenon is modelled by a backward propagating wave imposed on the forward propagating one. Starting with the TM Hertz vector, the only non-zero component of which reads

$\Pi_z(z,r,\varphi;t)=AJ_m(sr)\cos m\varphi\,\mathrm{e}^{j(\omega t-\beta z)}+AJ_m(sr)\cos m\varphi\,\mathrm{e}^{j(\omega t+\beta z)}$

The procedure involves taking the derivatives $\frac{\partial}{\partial r}$ $\frac{\partial}{\partial\varphi}$ first when required, before negating the variables i.e. taking $-r$ $-\varphi$, so as to model the backward propagating wave

TM (working with the roots of $J_m(sr)$), *improved reading in parallel to the original for* TE *below*

$$E_z=(k^2-\beta^2)\Pi_z$$

$$E_z=As^2\left(J_m(sr)\cos m\varphi\,\mathrm{e}^{j(\omega t-\beta z)}+AJ_m(sr)\cos m\varphi\,\mathrm{e}^{j(\omega t+\beta z)}\right) \qquad s^2=k^2-\beta^2$$

$$=As^2J_m(sr)\cos m\varphi\,\mathrm{e}^{j\omega t}\left(\mathrm{e}^{-j\beta z}+\mathrm{e}^{+j\beta z}\right) \qquad \beta=\frac{2\pi}{\Lambda}$$

$$=As^2J_m(sr)\cos m\varphi\,\mathrm{e}^{j\omega t}\,2\cos\frac{2\pi}{\Lambda}z \qquad J_m(sr_0)=0,\ \forall z,\varphi$$

Λ representing the measureable periodicity along the hollow cylinder as contrasted to the free wavelength $\lambda=c/f$ *also recalling that* $\lambda=c/(\omega/2\pi)=2\pi c/\omega=2\pi/k$, (*improved reading*)

$$E_r=-j\beta\frac{\partial\Pi_z}{\partial r}$$

$$E_r=-j\beta A\cos m\varphi\,\mathrm{e}^{j\omega t}\left(\frac{\partial}{\partial r}J_m(sr)\mathrm{e}^{-j\beta z}+\frac{\partial}{\partial r}J_m(sr)\mathrm{e}^{j\beta z}\right) \qquad \text{Taking derivatives}$$

$$=-j\beta A\cos m\varphi\,\mathrm{e}^{j\omega t}\left(sJ_m'(sr)\mathrm{e}^{-j\beta z}+sJ_m'(sr)\mathrm{e}^{j\beta z}\right) \qquad \text{Still taking derivatives}$$

$$=-j\beta A\cos m\varphi\,\mathrm{e}^{j\omega t}\left(sJ_m'(sr)\mathrm{e}^{-j\beta z}+sJ_m'(-sr)\mathrm{e}^{j\beta z}\right) \qquad \text{Negating } r \text{ to read } -r \text{ in 2nd term}$$

$$=-j\beta A\cos m\varphi\,\mathrm{e}^{j\omega t}\left(sJ_m'(sr)\mathrm{e}^{-j\beta z}-sJ_m'(sr)\mathrm{e}^{j\beta z}\right) \qquad \text{Using } J_m'(-sr)=-J_m'(sr) \text{ odd funct?}$$

This requires that m be even.

$$= -j\beta A s J'_m(sr)\cos m\varphi\, e^{j\omega t}\left(e^{-j\beta z} - e^{j\beta z}\right)$$

$$= j\beta A s J'_m(sr)\cos m\varphi\, e^{j\omega t}\left(-e^{-j\beta z} + e^{j\beta z}\right)$$

$$= j\beta A s J'_m(sr)\cos m\varphi\, e^{j\omega t}\, 2j\frac{\left(-e^{-j\beta z} + e^{j\beta z}\right)}{2j}$$

$$= -2\beta A s J'_m(sr)\cos m\varphi\, e^{j\omega t}\sin\beta z$$

$$= -2\beta A s J'_m(sr)\cos m\varphi\, e^{j\omega t}\sin\frac{2\pi}{\Lambda}z$$

$E_r = 0$ at the planes $z = \pm p\frac{\Lambda}{2}$ any integer p including zero

$J'_m(sr_0) \neq 0$, thus $E_r \neq 0$ at $z \neq \pm p\frac{\Lambda}{2}$ (OK, normal vector component)

$$E_\varphi = -j\frac{\beta}{r}\frac{\partial \Pi_z}{\partial \varphi}$$

$$E_\varphi = -j\frac{\beta}{r}AJ_m(sr)e^{j\omega t}\left(\frac{\partial}{\partial\varphi}\cos m\varphi\, e^{-j\beta z} + \frac{\partial}{\partial\varphi}\cos m\varphi\, e^{j\beta z}\right)$$ Taking derivatives

$$= -j\frac{\beta}{r}AJ_m(sr)e^{j\omega t}\left(-m\sin m\varphi\, e^{-j\beta z} - m\sin m\varphi\, e^{j\beta z}\right)$$ Still taking derivatives

$$= j\frac{\beta}{r}AJ_m(sr)e^{j\omega t}\left(m\sin m\varphi\, e^{-j\beta z} + m\sin m\varphi\, e^{j\beta z}\right)$$

$$= j\frac{\beta}{r}AJ_m(sr)e^{j\omega t}\left(m\sin m\varphi\, e^{-j\beta z} + m\sin(-m\varphi)\, e^{j\beta z}\right)$$ Negating φ to read $-\varphi$ in 2nd term

$$= j\frac{\beta}{r}AJ_m(sr)e^{j\omega t}\left(m\sin m\varphi\, e^{-j\beta z} - m\sin m\varphi\, e^{j\beta z}\right)$$ Using $\sin(-m\varphi) = -\sin m\varphi$

$$= j\frac{\beta}{r}AJ_m(sr)m\sin m\varphi\, e^{j\omega t}\left(e^{-j\beta z} - e^{j\beta z}\right)$$

$$= -j\frac{\beta}{r}AJ_m(sr)m\sin m\varphi\, e^{j\omega t}\left(-e^{-j\beta z} + e^{j\beta z}\right)$$

$$= -(2j)j\frac{\beta}{r}AJ_m(sr)m\sin m\varphi\, e^{j\omega t}\frac{\left(-e^{-j\beta z} + e^{j\beta z}\right)}{2j}$$

$$= 2\frac{\beta}{r}AJ_m(sr)m\sin m\varphi\, e^{j\omega t}\sin\beta z$$

$$= 2\frac{\beta}{r}AJ_m(sr)m\sin m\varphi\, e^{j\omega t}\sin\frac{2\pi}{\Lambda}z$$

$E_\varphi = 0$ at the planes $z = \pm p\frac{\Lambda}{2}$ any integer p including zero

TE (working with the roots of $J'_m(sr)$)

$E_z = 0$

$$E_r = -\frac{j\omega\mu}{r}\frac{\partial \Pi_z}{\partial \varphi}$$

$$E_r = -\frac{j\omega\mu}{r}AJ_m(sr)e^{j\omega t}\left(\frac{\partial}{\partial\varphi}\cos m\varphi\, e^{-j\beta z} + \frac{\partial}{\partial\varphi}\cos m\varphi\, e^{j\beta z}\right)$$ Taking derivatives

$$= -\frac{j\omega\mu}{r}AJ_m(sr)e^{j\omega t}\left(-m\sin m\varphi\, e^{-j\beta z} - m\sin m\varphi\, e^{j\beta z}\right)$$ Still taking derivatives

$$= \frac{j\omega\mu}{r} AJ_m(sr)\, e^{j\omega t}\left(m \sin m\varphi\, e^{-j\beta z} + m \sin m\varphi\, e^{j\beta z}\right)$$

$$= \frac{j\omega\mu}{r} AJ_m(sr)\, e^{j\omega t}\left(m \sin m\varphi\, e^{-j\beta z} + m \sin(-m\varphi)\, e^{j\beta z}\right)$$ Negating φ to read $-\varphi$ in 2nd term

$$= \frac{j\omega\mu}{r} AJ_m(sr)\, e^{j\omega t}\left(m \sin m\varphi\, e^{-j\beta z} - m \sin m\varphi\, e^{j\beta z}\right)$$ Using $\sin(-m\varphi) = -\sin m\varphi$

$$= \frac{j\omega\mu}{r} AJ_m(sr) m \sin m\varphi\, e^{j\omega t}\left(e^{-j\beta z} - e^{j\beta z}\right)$$

$$= -\frac{j\omega\mu}{r} AJ_m(sr) m \sin m\varphi\, e^{j\omega t}\left(-e^{-j\beta z} + e^{j\beta z}\right)$$

$$= -(2j)\frac{j\omega\mu}{r} AJ_m(sr) m \sin m\varphi\, e^{j\omega t}\, \frac{\left(-e^{-j\beta z} + e^{j\beta z}\right)}{2j}$$

$$= 2\frac{\omega\mu}{r} AJ_m(sr) m \sin m\varphi\, e^{j\omega t} \sin \beta z$$

$$= 2\frac{\omega\mu}{r} AJ_m(sr) m \sin m\varphi\, e^{j\omega t} \sin \frac{2\pi}{\Lambda} z$$

$E_r = 0$ at the planes $z = \pm p\frac{\Lambda}{2}$ any integer p including zero

$J_m(sr_0) \neq 0$, thus $E_r \neq 0$ at $z \neq \pm p\frac{\Lambda}{2}$ (OK, normal vector component)

$$E_\varphi = j\omega\mu \frac{\partial \Pi_z}{\partial r}$$

$$E_\varphi = j\omega\mu A \cos m\varphi\, e^{j\omega t}\left(\frac{\partial}{\partial r} J_m(sr)\, e^{-j\beta z} + \frac{\partial}{\partial r} J_m(sr)\, e^{j\beta z}\right)$$ Taking derivatives

$$= j\omega\mu A \cos m\varphi\, e^{j\omega t}\left(sJ'_m(sr)\, e^{-j\beta z} + sJ'_m(sr)\, e^{j\beta z}\right)$$ Still taking derivatives

$$= j\omega\mu A \cos m\varphi\, e^{j\omega t}\left(sJ'_m(sr)\, e^{-j\beta z} + sJ'_m(-sr)\, e^{j\beta z}\right)$$ Negating r to read $-r$ in 2nd term

$$= j\omega\mu A \cos m\varphi\, e^{j\omega t}\left(sJ'_m(sr)\, e^{-j\beta z} - sJ'_m(sr)\, e^{j\beta z}\right)$$ Using $J'_m(-sr) = -J'_m(sr)$ odd funct?

This requires that m be even.

$$= j\omega\mu A sJ'_m(sr) \cos m\varphi\, e^{j\omega t}\left(e^{-j\beta z} - e^{j\beta z}\right)$$

$$= -j\omega\mu A sJ'_m(sr) \cos m\varphi\, e^{j\omega t}\left(-e^{-j\beta z} + e^{j\beta z}\right)$$

$$= -j\omega\mu A sJ'_m(sr) \cos m\varphi\, e^{j\omega t}\, 2j \frac{\left(-e^{-j\beta z} + e^{j\beta z}\right)}{2j}$$

$$= 2\omega\mu A sJ'_m(sr) \cos m\varphi\, e^{j\omega t} \sin \beta z$$

$$= 2\omega\mu A sJ'_m(sr) \cos m\varphi\, e^{j\omega t} \sin \frac{2\pi}{\Lambda} z$$ $J'_m(sr_0) = 0$, $\forall z, \varphi$ roots of $J'_m(sr)$

$E_\varphi = 0$ at the planes $z = \pm p\frac{\Lambda}{2}$ any integer p including zero

For an odd m, $\sin\frac{2\pi}{\Lambda}z$ is replaced by $\frac{1}{j}\cos\frac{2\pi}{\Lambda}z$, $E_\varphi = 0$ and $E_r = 0$ occur at $z = \frac{\Lambda}{4} \pm p\frac{\Lambda}{2}$, the length $L = p\frac{\Lambda}{2}$ relates to Λ as shown

Letting $L = p\frac{\Lambda}{2}$ and using $\Lambda_{TM} = \frac{\lambda}{\sqrt{1-\left(\frac{a_{mn}\lambda}{2\pi r_0}\right)^2}}$ to write $L = \frac{p}{2}\frac{\lambda}{\sqrt{1-\left(\frac{a_{mn}\lambda}{2\pi r_0}\right)^2}}$ so as to arrive at the Eigenwavelength for the cavity resonator, a_{mn} the n-th root of $J_m(sr)$

$$\lambda_{TM,mnp} = \frac{2}{\sqrt{\frac{p^2}{L^2}+\left(\frac{a_{mn}}{\pi}\right)^2\frac{1}{r_0^2}}}$$

Likewise for TE, a'_{mn} the n-th root of $J'_m(sr)$

$$\lambda_{TE,mnp} = \frac{2}{\sqrt{\frac{p^2}{L^2}+\left(\frac{a'_{mn}}{\pi}\right)^2\frac{1}{r_0^2}}}$$

The indices $m\ n\ p$ correspond to the variables $\varphi\ r\ z$ and indicate, respectively, a mode of m angle dividing planes, involving the first n roots of the Bessel function, with $p = 0$ making the longest wavelength that aside from being longest is also independent of the cylinder length L.

§§§

Improved reading, a reminder, not a correction

Compare $\Lambda = 2\pi\frac{1}{\beta}$, measureable periodicity along the propagation direction for geometries other than the sphere, with the wavelength definition $\lambda = 2\pi\frac{1}{k}$. A reminder, the substitution $\rho = kr$, usually argument of Bessel functions, or $\rho = \beta r$ is dimensionless.

SPHERICAL HOLLOW CAVITY RESONATORS

$\omega = 2\pi f = 2\pi\frac{c}{\lambda}$ $\qquad \lambda = 2\pi\frac{c}{\omega} = 2\pi\frac{1}{\frac{\omega}{c}} = 2\pi\frac{1}{k}$ $\qquad$ where $k = \frac{\omega}{c} = \omega\sqrt{\varepsilon\mu}$

$k^2 = \omega^2\varepsilon\mu$ $\qquad \lambda = 2\pi\frac{r_0}{kr_0} = 2\pi\frac{r_0}{\rho_v}$ radius of the sphere coming into play

ρ_v is the v-th root of $J_n(\rho)$, TM_{mnv} mode; $\quad \rho_v$ is the v-th root of $J'_n(\rho)$, TE_{mnv} mode.

(Wrong as shown. Rather, ρ_v the roots of the transcendental equations for the boundary conditions).

Spherical waves

Ansatz $\Pi_r = R(r)S(\vartheta,\varphi)$ Separation of variables leads to $\frac{d^2R}{dr^2}+\left[k^2-\frac{n(n+1)}{r^2}\right]R = 0$ (note original symbol $\frac{d}{dr}$ for ordinary differentiation). Multiplying the latter with S yields $\frac{\partial^2\Pi_r}{\partial r^2}+\left[k^2-\frac{n(n+1)}{r^2}\right]\Pi_r = 0$ (note symbol $\frac{\partial}{\partial r}$ for partial differentiation), rewritten as $k^2\Pi_r+\frac{\partial^2\Pi_r}{\partial r^2} = \frac{n(n+1)}{r^2}$. For TM it happens to be $E_r = k^2\Pi_r+\frac{\partial^2\Pi_r}{\partial r^2}$ and thus equalling $= \frac{n(n+1)}{r^2}\Pi_r$. And for TE it happens to be $H_r = k^2\Pi_r+\frac{\partial^2\Pi_r}{\partial r^2}$ and thus equalling $= \frac{n(n+1)}{r^2}\Pi_r$

Resonators

Property

Assuming ideal electric conductivity $\sigma \to \infty$ for cavity resonator of arbitrary shape and geometry, enclosing perfect isolator of material constants ε and μ, the following holds, Cf. **Addendum**.

1.) The angular Eigenfrequencies ω_i associated with Eigenvalues $k_i^2 = \varepsilon\mu\omega_i^2$ represent a sequence of distinct discrete real numbers. Solutions to the boundary problem are possible only for those Eigenvalues.

2.) Solutions associated with the various Eigenvalues are orthogonal to each other, i.e. $\int_V \mu \mathbf{H}_i \mathbf{H}_k^* dV = 0$ and $\int_V \varepsilon \mathbf{E}_i \mathbf{E}_k^* dV = 0$

3.) Slight imperfections of the cavity's geometry may be modelled by the relative deviation in frequency given by the estimation $\left|\dfrac{\omega_k - \omega_i}{\omega_i}\right| \approx \dfrac{\left|\int_{\Delta V} (\mu |\mathbf{H}_i|^2 - \varepsilon |\mathbf{E}_i|^2) dV\right|}{\int_{V'} (\mu |\mathbf{H}_i|^2 + \varepsilon |\mathbf{E}_i|^2) dV}$ ω_i representing the frequency for the perfect geometry.

Addendum

Expression	Conj.	Multiplier
$curl\ \mathbf{H}_i = j\omega_i\varepsilon\ \mathbf{E}_i$	$curl\ \mathbf{H}_i^* = -j\omega_i\varepsilon\ \mathbf{E}_i^*$	$-\mathbf{E}_i$
$curl\ \mathbf{E}_i = -j\omega_i\mu\ \mathbf{H}_i$	$curl\ \mathbf{E}_i^* = j\omega_i\mu\ \mathbf{H}_i^*$	$\mathbf{H}_i^*$
$curl\ \mathbf{H}_k^* = -j\omega_k\varepsilon\ \mathbf{E}_k^*$	$curl\ \mathbf{H}_k = j\omega_k\varepsilon\ \mathbf{E}_k$	
$curl\ \mathbf{E}_k^* = j\omega_k\mu\ \mathbf{H}_k^*$	$curl\ \mathbf{E}_k = -j\omega_k\mu\ \mathbf{H}_k$	

$(curl\ \mathbf{H}_i^* = -j\omega_i\varepsilon\ \mathbf{E}_i^*)(-\mathbf{E}_i) \to -\mathbf{E}_i\, curl\ \mathbf{H}_i^* = j\omega_i\varepsilon\ \mathbf{E}_i\mathbf{E}_i^*$

$(curl\ \mathbf{E}_i = -j\omega_i\mu\ \mathbf{H}_i)(\mathbf{H}_i^*) \to \mathbf{H}_i^*\, curl\ \mathbf{E}_i = -j\omega_i\mu\ \mathbf{H}_i\mathbf{H}_i^*$

Adding

$$\mathbf{H}_i^*\, curl\ \mathbf{E}_i - \mathbf{E}_i\, curl\ \mathbf{H}_i^* = -j\omega_i\mu\ \mathbf{H}_i\mathbf{H}_i^* + j\omega_i\varepsilon\ \mathbf{E}_i\mathbf{E}_i^*$$

Using

$div(\mathbf{u}\times\mathbf{v}) = \mathbf{v}\ curl\ \mathbf{u} - \mathbf{u}\ curl\ \mathbf{v}$

$\mathbf{H}_i^*\, curl\ \mathbf{E}_i - \mathbf{E}_i\, curl\ \mathbf{H}_i^* = div(\mathbf{E}_i \times \mathbf{H}_i^*)$

Using $curl\ \mathbf{H}_i = j\omega_i\varepsilon\ \mathbf{E}_i$

$$\mathbf{E}_i = \frac{1}{j\omega_i\varepsilon} curl\ \mathbf{H}_i \text{ and } \mathbf{E}_i^* = \frac{-1}{j\omega_i\varepsilon} curl\ \mathbf{H}_i^*$$

$$\mathbf{E}_i\mathbf{E}_i^* = \frac{1}{j\omega_i\varepsilon} curl\ \mathbf{H}_i \frac{-1}{j\omega_i\varepsilon} curl\ \mathbf{H}_i^* =$$

$$\frac{1}{(\omega_i\varepsilon)^2} curl\ \mathbf{H}_i curl\ \mathbf{H}_i^*$$

$$div(\mathbf{E}_i \times \mathbf{H}_i^*) =$$

$$-j\omega_i\mu\ \mathbf{H}_i\mathbf{H}_i^* + j\omega_i\varepsilon \frac{1}{(\omega_i\varepsilon)^2} curl\ \mathbf{H}_i curl\ \mathbf{H}_i^*$$

$$j\omega_i\varepsilon\ div(\mathbf{E}_i \times \mathbf{H}_i^*) =$$

$$-j\omega_i\mu\ j\omega_i\varepsilon\ \mathbf{H}_i\mathbf{H}_i^* + (j\omega_i\varepsilon)^2 \frac{1}{(\omega_i\varepsilon)^2} curl\ \mathbf{H}_i curl\ \mathbf{H}_i^*$$

$$j\omega_i\varepsilon\ div(\mathbf{E}_i \times \mathbf{H}_i^*) = \varepsilon\mu\ \omega_i^2\ \mathbf{H}_i\mathbf{H}_i^* - curl\ \mathbf{H}_i curl\ \mathbf{H}_i^*$$

$$j\omega_i\varepsilon\ div(\mathbf{E}_i \times \mathbf{H}_i^*) = \varepsilon\mu\ \omega_i^2\ \mathbf{H}_i\mathbf{H}_i^* - (curl\ \mathbf{H}_i)(curl\ \mathbf{H}_i)^*$$

$$j\omega_i\varepsilon\ div(\mathbf{E}_i \times \mathbf{H}_i^*) = \varepsilon\mu\ \omega_i^2\ |\mathbf{H}_i|^2 - |curl\ \mathbf{H}_i|^2$$

$$\varepsilon\mu\ \omega_i^2\ |\mathbf{H}_i|^2 = |curl\ \mathbf{H}_i|^2 + j\omega_i\varepsilon\ div(\mathbf{E}_i \times \mathbf{H}_i^*)$$

$$\varepsilon\mu\ \omega_i^2 = \frac{\int_V |curl\ \mathbf{H}_i|^2\ dV + \int_V j\omega_i\varepsilon\ div(\mathbf{E}_i \times \mathbf{H}_i^*) dV}{\int_V |\mathbf{H}_i|^2\ dV}$$

applying Gauß and using the boundary condition $\mathbf{n} \times \mathbf{E}_i = 0$ (for $\sigma \to \infty$ perfect metal)

$$\int_V div(\mathbf{E}_i \times \mathbf{H}_i^*) dV = \oint_A (\mathbf{E}_i \times \mathbf{H}_i^*) d\mathbf{A}$$

$$= \oint_A (\mathbf{E}_i \times \mathbf{H}_i^*)\mathbf{n} dA = \oint_A (\mathbf{n} \times \mathbf{E}_i)\mathbf{H}_i^* dA = 0$$

showing

$$k_i^2 = \varepsilon\mu\ \omega_i^2 = \frac{\int_V |curl\ \mathbf{H}_i|^2\ dV}{\int_V |\mathbf{H}_i|^2\ dV} \geq 0$$

The equal sign applies only for the trivial case $\mathbf{H}_i \equiv 0$. Solutions to the boundary problem thus possible only for distinct discrete angular (positive real) Eigenfrequencies.

Expression	Conj.	Multiplier
$curl\ \mathbf{H}_i = j\omega_i\varepsilon\ \mathbf{E}_i$	$curl\ \mathbf{H}_i^* = -j\omega_i\varepsilon\ \mathbf{E}_i^*$	$-\mathbf{E}_k^*$, a
$curl\ \mathbf{E}_i = -j\omega_i\mu\ \mathbf{H}_i$	$curl\ \mathbf{E}_i^* = j\omega_i\mu\ \mathbf{H}_i^*$	$\mathbf{H}_k^*$, A
$curl\ \mathbf{H}_k^* = -j\omega_k\varepsilon\ \mathbf{E}_k^*$	$curl\ \mathbf{H}_k = j\omega_k\varepsilon\ \mathbf{E}_k$	$-\mathbf{E}_i$, b
$curl\ \mathbf{E}_k^* = j\omega_k\mu\ \mathbf{H}_k^*$	$curl\ \mathbf{E}_k = -j\omega_k\mu\ \mathbf{H}_k$	$\mathbf{H}_i$, B

On carrying out the implied multiplications,
Adding eq. b to eq. A

$div(\mathbf{E}_i \times \mathbf{H}_k^*) = -j\ \omega_i\ \mu\ \mathbf{H}_i\mathbf{H}_k^* + j\ \omega_k\varepsilon\ \mathbf{E}_i\mathbf{E}_k^*$ Adding eq. a to eq. B

$div(\mathbf{E}_k^* \times \mathbf{H}_i) = j\ \omega_k\mu\ \mathbf{H}_i\mathbf{H}_k^* - j\ \omega_i\ \varepsilon\ \mathbf{E}_i\mathbf{E}_k^*$

On converting the volume to surface integrals and using $\mathbf{n} \times \mathbf{E}_i = 0$ and $\mathbf{n} \times \mathbf{E}_k = 0$, the surface integrals vanish, leaving

$$0=-\omega_i \int_V \mu\, \mathbf{H}_i \mathbf{H}_k^* dV + \omega_k \int_V \varepsilon\, \mathbf{E}_i \mathbf{E}_k^* dV$$

$$0=\omega_k \int_V \mu\, \mathbf{H}_i \mathbf{H}_k^* dV - \omega_i \int_V \varepsilon\, \mathbf{E}_i \mathbf{E}_k^* dV$$

$$\begin{pmatrix} -\omega_i & \omega_k \\ \omega_k & -\omega_i \end{pmatrix} \begin{pmatrix} \int_V \mu\, \mathbf{H}_i \mathbf{H}_k^* dV \\ \int_V \varepsilon\, \mathbf{E}_i \mathbf{E}_k^* dV \end{pmatrix} = \begin{pmatrix} 0 \\ 0 \end{pmatrix}$$

Non-trivial solutions for zero determinant $\omega_i^2 - \omega_k^2 = 0$, i.e. equal Eigenvalues $k_i^2 = k_k^2$, i.e. equal indices $i = k$ in which case $\int_V \mu\, |\mathbf{H}_i|^2\, dV = \int_V \varepsilon\, |\mathbf{E}_i|^2\, dV$ telling of the equivalence between magnetic and electric energies (equivalent in time average).

Interesting when $\omega_i^2 \neq \omega_k^2$ where the solutions associated with different Eigenfrequencies are orthogonal to each other, in particular

$\int_V \mu \mathbf{H}_i \mathbf{H}_k^* dV = 0$ and $\int_V \varepsilon \mathbf{E}_i \mathbf{E}_k^* dV = 0$

Modelling imperfection, Take 2

Repeating

Adding eq. b to eq. A

$div(\mathbf{E}_i \times \mathbf{H}_k^*) = -j\, \omega_i\, \mu\, \mathbf{H}_i \mathbf{H}_k^* + j\, \omega_k \varepsilon\, \mathbf{E}_i \mathbf{E}_k^*$ Adding eq. a to eq. B

$div(\mathbf{E}_k^* \times \mathbf{H}_i) = j\, \omega_k \mu\, \mathbf{H}_i \mathbf{H}_k^* - j\, \omega_i\, \varepsilon\, \mathbf{E}_i \mathbf{E}_k^*$

$$\begin{aligned} div(\mathbf{E}_i \times \mathbf{H}_k^*) + div(\mathbf{E}_k^* \times \mathbf{H}_i) = &\\ &+ j\omega_k \varepsilon\, \mathbf{E}_i \mathbf{E}_k^* - j\omega_i \mu\, \mathbf{H}_i \mathbf{H}_k^* \\ &- j\omega_i \varepsilon\, \mathbf{E}_i \mathbf{E}_k^* + j\omega_k \mu\, \mathbf{H}_i \mathbf{H}_k^* \end{aligned}$$

$$\begin{aligned} div(\mathbf{E}_i \times \mathbf{H}_k^*) + div(\mathbf{E}_k^* \times \mathbf{H}_i) = &\\ &+ j(\omega_k - \omega_i)\varepsilon\, \mathbf{E}_i \mathbf{E}_k^* + j(\omega_k - \omega_i)\mu\, \mathbf{H}_i \mathbf{H}_k^* \end{aligned}$$

$$\begin{aligned} div(\mathbf{E}_i \times \mathbf{H}_k^*) + div(\mathbf{E}_k^* \times \mathbf{H}_i) = &\\ &+ j(\omega_k - \omega_i)(\varepsilon\, \mathbf{E}_i \mathbf{E}_k^* + \mu\, \mathbf{H}_i \mathbf{H}_k^*) \end{aligned}$$

$$\begin{aligned} \int_{V'} div(\mathbf{E}_i \times \mathbf{H}_k^*) dV + \int_{V'} div(\mathbf{E}_k^* \times \mathbf{H}_i) dV = &\\ &+ j(\omega_k - \omega_i) \int_{V'} (\varepsilon\, \mathbf{E}_i \mathbf{E}_k^* + \mu\, \mathbf{H}_i \mathbf{H}_k^*) dV \end{aligned}$$

$$\begin{aligned} \oint_{A'} (\mathbf{E}_i \times \mathbf{H}_k^*) d\mathbf{A} + \oint_{A'} (\mathbf{E}_k^* \times \mathbf{H}_i) d\mathbf{A} = &\\ &+ j(\omega_k - \omega_i) \int_{V'} (\varepsilon\, \mathbf{E}_i \mathbf{E}_k^* + \mu\, \mathbf{H}_i \mathbf{H}_k^*) dV \end{aligned}$$

Looking @ the 1st surface integral on the LHS $(\mathbf{E}_i \times \mathbf{H}_k^*) d\mathbf{A} = (\mathbf{n} \times \mathbf{E}_i) \mathbf{H}_k^* dA = 0$

As for the 2nd one

$\oint_{A'} (\mathbf{E}_k^* \times \mathbf{H}_i) d\mathbf{A} = \oint_{A} (\mathbf{E}_k^* \times \mathbf{H}_i) d\mathbf{A} - \oint_{\Delta A} (\mathbf{E}_k^* \times \mathbf{H}_i) d\mathbf{A}$ the 1st term of which is zero

$(\mathbf{E}_k^* \times \mathbf{H}_i) d\mathbf{A} = (\mathbf{n} \times \mathbf{E}_k^*) \mathbf{H}_i\, dA = 0$

leaving the LHS as

$$\oint_{A'} (\mathbf{E}_k^* \times \mathbf{H}_i) d\mathbf{A} = -\oint_{\Delta A} (\mathbf{E}_k^* \times \mathbf{H}_i) d\mathbf{A}$$

Thus

$$-\oint_{\Delta A} (\mathbf{E}_k^* \times \mathbf{H}_i) d\mathbf{A} =$$

$$+ j(\omega_k - \omega_i) \int_{V'} (\varepsilon\, \mathbf{E}_i \mathbf{E}_k^* + \mu\, \mathbf{H}_i \mathbf{H}_k^*) dV$$

$$\omega_k - \omega_i = j \frac{\oint_{\Delta A} (\mathbf{E}_k^* \times \mathbf{H}_i) d\mathbf{A}}{\int_{V'} (\varepsilon\, \mathbf{E}_i \mathbf{E}_k^* + \mu\, \mathbf{H}_i \mathbf{H}_k^*) dV}$$

$\mathbf{E}_k$ and $\mathbf{H}_k$ for the imperfect case are not known, may however be modelled by the approximations $\mathbf{E}_k \approx \mathbf{E}_i$ and $\mathbf{H}_k \approx \mathbf{H}_i$

$$\omega_k - \omega_i \approx j \frac{\oint_{\Delta A} (\mathbf{E}_i^* \times \mathbf{H}_i) d\mathbf{A}}{\int_{V'} (\varepsilon\, \mathbf{E}_i \mathbf{E}_i^* + \mu\, \mathbf{H}_i \mathbf{H}_i^*) dV}$$

$$\omega_k - \omega_i \approx j \frac{-\oint_{\Delta A} (\mathbf{H}_i \times \mathbf{E}_i^*) d\mathbf{A}}{\int_{V'} (\varepsilon |\mathbf{E}_i|^2 + \mu |\mathbf{H}_i|^2) dV}$$

By the relationship

$$\oint_{\Delta A} (\mathbf{H}_i \times \mathbf{E}_i^*) d\mathbf{A} = j\omega_i \int_{\Delta V} (\varepsilon\, \mathbf{E}_i \mathbf{E}_i^* - \mu\, \mathbf{H}_i \mathbf{H}_i^*) dV$$

$$\left| \frac{\omega_k - \omega_i}{\omega_i} \right| \approx \frac{\left| \int_{\Delta V} (\mu |\mathbf{H}_i|^2 - \varepsilon |\mathbf{E}_i|^2) dV \right|}{\int_{V'} (\mu |\mathbf{H}_i|^2 + \varepsilon |\mathbf{E}_i|^2) dV}$$

where V' may be approximated by $V \approx V'$ when appropriate.

§§§

$\lambda/2$ Dipole Arrays'Closed-form Expressions

<table>
<tr><td></td><td>Vertically spaced array of $2m+1$ vertical $\lambda/2$ dipoles (or primary antennas)</td><td>Horizontally spaced array of $2m+1$ vertical $\lambda/2$ dipoles (or primary antennas)</td></tr>
<tr><td>Generic expression for $E_{\theta\nu}$</td><td colspan="2">$E_{\theta\nu} = magnitude \times e^{j\omega t} \times e^{-j\mathbf{kr}}$
(θ-component of the electric vector due to the ν-th primary antenna)</td></tr>
<tr><td>Far field θ-component</td><td colspan="2">$E_{\theta\nu} = \frac{60}{r_\nu} I \frac{\cos(\frac{\pi}{2}\cos\theta_\nu)}{\sin\theta_\nu} e^{j\omega\left(t-\frac{r_\nu}{c}\right)}$
$E_{\theta\nu} = \frac{60}{r_\nu} I \frac{\cos(\frac{\pi}{2}\cos\theta_\nu)}{\sin\theta_\nu} e^{j\omega\left(t-\frac{r_0-d_\nu}{c}\right)}$
$E_{\theta\nu} = \frac{60}{r_\nu} I \frac{\cos(\frac{\pi}{2}\cos\theta_\nu)}{\sin\theta_\nu} e^{j\omega\left(t-\frac{r_0}{c}\right)} e^{j\omega\frac{d_\nu}{c}}$
$E_{\theta\nu} = \frac{60}{r_\nu} I \frac{\cos(\frac{\pi}{2}\cos\theta_\nu)}{\sin\theta_\nu} e^{j\omega\left(t-\frac{r_0}{c}\right)} e^{j\frac{2\pi}{\lambda}d_\nu}$
(The phasing term $e^{j\omega\left(t-\frac{r_0}{c}\right)}$ with unity magnitude common to all ν would be of little interest; The actual expression would have read $\approx j60\Omega\cdots$, however the Ohmic symbol Ω should not be part of the expression, nor would j play a role in the magnitude computation.)</td></tr>
<tr><td>${}^0\mathbf{u}_P$</td><td colspan="2">${}^0\mathbf{u}_P = \begin{pmatrix} \sin\theta_0\cos\varphi_0 \\ \sin\theta_0\sin\varphi_0 \\ \cos\theta_0 \\ 0 \end{pmatrix}$</td></tr>
<tr><td>${}^0\mathbf{u}_{y_\nu}$</td><td></td><td>${}^0\mathbf{u}_{y_\nu} = \begin{pmatrix} 0 \\ 1 \\ 0 \\ 0 \end{pmatrix}$</td></tr>
<tr><td>${}^0\mathbf{u}_{z_\nu}$</td><td>${}^0\mathbf{u}_{z_\nu} = \begin{pmatrix} 0 \\ 0 \\ 1 \\ 0 \end{pmatrix}$</td><td></td></tr>
</table>

${}^0\mathbf{u}_P\,{}^0\mathbf{u}_{y_\nu}$		$\sin\theta_0\sin\varphi_0$ (Projection of ${}^0\mathbf{u}_{y_\nu}$ onto the direction of propagation)
${}^0\mathbf{u}_P\,{}^0\mathbf{u}_{z_\nu}$	$\cos\theta_0$ (Projection of ${}^0\mathbf{u}_{z_\nu}$ onto the direction of propagation)	
d_ν	$d_\nu = \nu\frac{\lambda}{2}\cos\theta_0$ (Path length difference approximation at the ν-th $\lambda/2$ separation)	$d_\nu = \nu\frac{\lambda}{2}\sin\theta_0\sin\varphi_0$ (Path length difference approximation at the ν-th $\lambda/2$ separation)
d_ν	$d_\nu = r_0 - r_\nu$ $r_\nu = \sqrt{(\nu\frac{\lambda}{2})^2 - 2(\nu\frac{\lambda}{2})r_0\cos\gamma + r_0^2}$ $\cos\gamma = \operatorname{sgn}\nu\cos\theta_0$ (original definition)	$d_\nu = r_0 - r_\nu$ $r_\nu = \sqrt{(\nu\frac{\lambda}{2})^2 - 2(\nu\frac{\lambda}{2})r_0\cos\gamma + r_0^2}$ $\cos\gamma = \operatorname{sgn}\nu\sin\theta_0\sin\varphi_0$ (original definition)
$E_\theta / \frac{60I}{r_0}$	$E_\theta / \frac{60I}{r_0} =$ $\frac{\cos(\frac{\pi}{2}\cos\theta_0)}{\sin\theta_0} e^{j\omega\left(t-\frac{r_0}{c}\right)} \sum_{\nu=-m}^{m}\left(e^{j\pi\cos\theta_0}\right)^\nu$	$E_\theta / \frac{60I}{r_0} =$ $\frac{\cos(\frac{\pi}{2}\cos\theta_0)}{\sin\theta_0} e^{j\omega\left(t-\frac{r_0}{c}\right)} \sum_{\nu=-m}^{m}\left(e^{j\pi\sin\theta_0\sin\varphi_0}\right)^\nu$
$\lvert E_\theta / \frac{60I}{r_0}\rvert$	$\frac{\cos(\frac{\pi}{2}\cos\theta_0)}{\sin\theta_0} \lvert \sum_{\nu=-m}^{m}\left(e^{j\pi\cos\theta_0}\right)^\nu \rvert$	$\frac{\cos(\frac{\pi}{2}\cos\theta_0)}{\sin\theta_0} \lvert \sum_{\nu=-m}^{m}\left(e^{j\pi\sin\theta_0\sin\varphi_0}\right)^\nu \rvert$
$\lvert E_\theta / \frac{60I}{r_0}\rvert$	$\frac{\cos(\frac{\pi}{2}\cos\theta_0)}{\sin\theta_0}\frac{\sin(\frac{n\pi}{2}\cos\theta_0)}{\sin(\frac{\pi}{2}\cos\theta_0)}$ $0 \le \theta_0 \le \pi$	$\frac{\cos(\frac{\pi}{2}\cos\theta_0)}{\sin\theta_0}\frac{\sin(\frac{n\pi}{2}\sin\theta_0\sin\varphi_0)}{\sin(\frac{\pi}{2}\sin\theta_0\sin\varphi_0)}$ $0 \le \theta_0 \le \pi$, $0 \le \varphi_0 < 2\pi$

Letting e.g. $x \equiv e^{j\pi\sin\theta_0\sin\varphi_0}$ so as to check $\sum_{\nu=-m}^{m} x^\nu$, the index substitution $\mu = \nu + m$ yields $\sum_{\mu=0}^{2m} x^{\mu-m} = x^{-m}\sum_{\mu=0}^{2m} x^\mu$. The prefactor with unity magnitude is not entering the results of current interest. The following makes use of the formula $\sum_{\mu=0}^{2m} x^\mu = \frac{1-x^{2m+1}}{1-x}$ where it's safe to apply superposition for homogeneous media, Cf. below[3]. Complex conjugation would be involved in working out the magnitude of interest. Letting $r_\nu \approx r_0$, $\theta_\nu \approx \theta_0$, $\varphi_\nu \approx \varphi_0$, $\forall\nu$ for the farfield, also $n = 2m+1$, for a horizontally spaced array of vertical $\lambda/2$ dipoles or primary antennas

[3] It is known that [Bronstein, Semendjajew] $\sum_{\nu=0}^{\infty} x^\nu = \frac{1}{1-x}$, $|x|<1$, thus $\sum_{\nu=0}^{m} x^\nu = \frac{1}{1-x} - \sum_{\nu=m+1}^{\infty} x^\nu$. Shifting the index $\mu = \nu - (m+1)$ to write $\sum_{\nu=0}^{m} x^\nu = \frac{1}{1-x} - \sum_{\mu=0}^{\infty} x^{\mu+(m+1)} = \frac{1}{1-x} - x^{m+1}\sum_{\mu=0}^{\infty} x^\mu = \frac{1}{1-x} - x^{m+1}\frac{1}{1-x}$ and arrive at $\sum_{\nu=0}^{m} x^\nu = \frac{1-x^{m+1}}{1-x}$.

$$|^0\mathbf{E}_P| = \frac{60}{r_0} I \frac{\cos(\frac{\pi}{2}\cos\theta_0)}{\sin\theta_0} \frac{\sin\left(\frac{n\pi}{2}\sin\theta_0\sin\varphi_0\right)}{\sin\left(\frac{\pi}{2}\sin\theta_0\sin\varphi_0\right)} \quad 0 \le \theta_0 \le \pi\,,\ 0 \le \varphi_0 < 2\pi$$

and for a vertically spaced array of vertical $\lambda/2$ dipoles or primary antennas

$$|^0\mathbf{E}_P| = \frac{60}{r_0} I \frac{\cos(\frac{\pi}{2}\cos\theta_0)}{\sin\theta_0} \frac{\sin\left(\frac{n\pi}{2}\cos\theta_0\right)}{\sin\left(\frac{\pi}{2}\cos\theta_0\right)} \quad 0 \le \theta_0 \le \pi$$

known as E-plane patterns.

Addendum

Elementary dipole $l << \lambda$ recalled

Procedure $\mathbf{H} = curl\ j\omega\varepsilon\mathbf{\Pi}_e$ $\mathbf{E} = curl\ curl\ \mathbf{\Pi}_e$ $\mathbf{A} = j\omega\varepsilon\mathbf{\Pi}_e$ $\mathbf{A}(\mathbf{r}_P, t) = \frac{1}{4\pi}\int_V \frac{\mathbf{J}(\mathbf{r}_Q, t - \frac{r_{PQ}}{c})}{r_{PQ}} dV$, given $\mathbf{J}$

$$\mathbf{E} = \begin{pmatrix} E_r \\ E_\theta \\ E_\varphi \end{pmatrix} = \begin{pmatrix} \frac{I_0 l}{4\pi}\left[\sqrt{\frac{\mu_0}{\varepsilon_0}}\cdot\frac{2}{r^2} - \frac{2j}{\omega\varepsilon_0 r^3}\right]\cos\theta\cdot e^{j(\omega t - kr)} \\ \frac{I_0 l}{4\pi}\left[\frac{j\omega\mu_0}{r} + \sqrt{\frac{\mu_0}{\varepsilon_0}}\cdot\frac{1}{r^2} - \frac{j}{\omega\varepsilon_0 r^3}\right]\sin\theta\cdot e^{j(\omega t - kr)} \\ 0 \end{pmatrix}$$

Take 1

$$\mathbf{E} = \begin{pmatrix} E_r \\ E_\theta \\ E_\varphi \end{pmatrix} = \frac{I_0 l}{4\pi}\sqrt{\frac{\mu_0}{\varepsilon_0}} \begin{pmatrix} \left(\frac{2}{r^2} - \sqrt{\frac{\varepsilon_0}{\mu_0}}\frac{2j}{\omega\varepsilon_0 r^3}\right)\cos\theta\cdot e^{j(\omega t - kr)} \\ \left(\sqrt{\frac{\varepsilon_0}{\mu_0}}\omega\mu_0\frac{j}{r} + \frac{1}{r^2} - \sqrt{\frac{\varepsilon_0}{\mu_0}}\frac{j}{\omega\varepsilon_0 r^3}\right)\sin\theta\cdot e^{j(\omega t - kr)} \\ 0 \end{pmatrix}$$

Take 2

$$\mathbf{E} = \begin{pmatrix} E_r \\ E_\theta \\ E_\varphi \end{pmatrix} = \frac{I_0 l}{4\pi}\sqrt{\frac{\mu_0}{\varepsilon_0}} \begin{pmatrix} \left(\frac{2}{r^2} - \frac{1}{\omega\sqrt{\mu_0\varepsilon_0}}\frac{2j}{r^3}\right)\cos\theta\cdot e^{j(\omega t - kr)} \\ \left(\omega\sqrt{\mu_0\varepsilon_0}\frac{j}{r} + \frac{1}{r^2} - \frac{1}{\omega\sqrt{\mu_0\varepsilon_0}}\frac{j}{r^3}\right)\sin\theta\cdot e^{j(\omega t - kr)} \\ 0 \end{pmatrix}$$

Take 3

$$\mathbf{E} = \begin{pmatrix} E_r \\ E_\theta \\ E_\varphi \end{pmatrix} = \frac{I_0 l}{4\pi}\sqrt{\frac{\mu_0}{\varepsilon_0}} \begin{pmatrix} \left(\frac{2}{r^2} - \frac{1}{k}\frac{2j}{r^3}\right)\cos\theta \cdot e^{j(\omega t - kr)} \\ \left(k\frac{j}{r} + \frac{1}{r^2} - \frac{1}{k}\frac{j}{r^3}\right)\sin\theta \cdot e^{j(\omega t - kr)} \\ 0 \end{pmatrix} \text{ where } k^2 = \varepsilon_0\mu_0\omega^2 = -j\omega\mu_0 . j\omega\varepsilon_0$$

$$\mathbf{H} = \begin{pmatrix} H_r \\ H_\theta \\ H_\varphi \end{pmatrix} = \frac{I_0 l}{4\pi} \begin{pmatrix} 0 \\ 0 \\ \left(\frac{jk}{r} + \frac{1}{r^2}\right)\sin\theta \cdot e^{j(\omega t - kr)} \end{pmatrix}$$

Take 4

From the θ-component for the electric field

$$\frac{I_0 l}{4\pi}\sqrt{\frac{\mu_0}{\varepsilon_0}}\left(k\frac{j}{r} + \frac{1}{r^2} - \frac{1}{k}\frac{j}{r^3}\right)$$

writing

$$\frac{I_0 l}{4\pi}\sqrt{\frac{\mu_0}{\varepsilon_0}}\left(\frac{k^2}{k}\frac{j}{r} + \frac{k^2}{k^2}\frac{1}{r^2} - \frac{k^2}{k^3}\frac{j}{r^3}\right) = \frac{I_0 l}{4\pi}\sqrt{\frac{\mu_0}{\varepsilon_0}}k^2\left(\frac{j}{(kr)} + \frac{1}{(kr)^2} - \frac{j}{(kr)^3}\right)$$

$$= \frac{I_0 l}{4\pi}\sqrt{\frac{\mu_0}{\varepsilon_0}}k^2\frac{j}{j}\left(\frac{j}{(kr)} + \frac{1}{(kr)^2} - \frac{j}{(kr)^3}\right) = \frac{I_0 l}{4\pi}\sqrt{\frac{\mu_0}{\varepsilon_0}}k^2\frac{1}{j}\left(-\frac{1}{(kr)} + \frac{j}{(kr)^2} + \frac{1}{(kr)^3}\right)$$

thus $E_\theta = \frac{I_0 l}{4\pi}\sqrt{\frac{\mu_0}{\varepsilon_0}}k^2\frac{1}{j}\left(-\frac{1}{(kr)} + \frac{j}{(kr)^2} + \frac{1}{(kr)^3}\right)\sin\theta \cdot e^{j(\omega t - kr)}$

$$\mathbf{E} = \begin{pmatrix} E_r \\ E_\theta \\ E_\varphi \end{pmatrix} = \frac{I_0 l}{4\pi}\sqrt{\frac{\mu_0}{\varepsilon_0}}k^2\frac{1}{j} \begin{pmatrix} \left(\frac{2j}{(kr)^2} + \frac{2}{(kr)^3}\right)\cos\theta \cdot e^{j(\omega t - kr)} \\ \left(-\frac{1}{(kr)} + \frac{j}{(kr)^2} + \frac{1}{(kr)^3}\right)\sin\theta \cdot e^{j(\omega t - kr)} \\ 0 \end{pmatrix}$$

$$\mathbf{H} = \begin{pmatrix} H_r \\ H_\theta \\ H_\varphi \end{pmatrix} = \frac{I_0 l}{4\pi}k^2\frac{1}{j} \begin{pmatrix} 0 \\ 0 \\ \left(-\frac{1}{(kr)} + \frac{j}{(kr)^2}\right)\sin\theta \cdot e^{j(\omega t - kr)} \end{pmatrix}$$

The ratio between the electric/magnetic amplitudes may be seen as $\sqrt{\frac{\mu_0}{\varepsilon_0}} \approx 120\pi\Omega$ (which makes sense) known as the field resistance for plane waves in free space.

Returning to the initial expression to consider the farfield $E_\theta = \frac{I_0 l}{4\pi} j\omega\mu_0 \cdot \frac{1}{r}\sin\theta \cdot e^{j(\omega t - kr)}$, where substituting $\omega = 2\pi\frac{c}{\lambda}$ and $c = \frac{1}{\sqrt{\varepsilon_0\mu_0}}$ leads to

$$E_\theta = \frac{I_0 l}{4\pi} j\mu_0 \cdot \omega \cdot (\frac{1}{r}\sin\theta \cdot e^{j(\omega t - kr)}) \qquad = \frac{I_0 l}{4\pi} j\mu_0 \cdot 2\pi\frac{c}{\lambda} \cdot (\frac{1}{r}\sin\theta \cdot e^{j(\omega t - kr)})$$

$$= j\frac{1}{2}\mu_0 c(\frac{l}{\lambda})I_0 \cdot (\frac{1}{r}\sin\theta \cdot e^{j(\omega t - kr)}) \qquad = j\frac{1}{2}\mu_0\frac{1}{\sqrt{\varepsilon_0\mu_0}}(\frac{l}{\lambda})I_0 \cdot (\frac{1}{r}\sin\theta \cdot e^{j(\omega t - kr)})$$

$$= j\frac{1}{2}\sqrt{\frac{\mu_0}{\varepsilon_0}}(\frac{l}{\lambda})I_0 \cdot (\frac{1}{r}\sin\theta \cdot e^{j(\omega t - kr)}) \approx j60\pi\Omega(\frac{l}{\lambda})I_0 \cdot (\frac{1}{r}\sin\theta \cdot e^{j(\omega t - kr)}) \text{ where } (\frac{l}{\lambda}) << 1 \text{ is important}$$

Antennas of multiple $\lambda/2$ lengths may be modelled as linear combinations of dipoles.

The field for the spherical antenna is complex involving both TE & TM modes of all orders m and n, but for a generic comparison, looking at one (Eigen)frequency, assuming $\frac{\partial}{\partial\varphi} \equiv 0$ i.e. $m = 0$ and proceeding for TM only

$$\mathbf{E} = \begin{pmatrix} E_r \\ E_\theta \\ E_\varphi \end{pmatrix} = \begin{pmatrix} \sum_{n,m,\omega} a_{0n} k^2 \frac{n(n+1)}{(kr)^{\frac{3}{2}}} H^{(2)}_{n+\frac{1}{2}}(kr) P_n(\cos\theta) \cdot e^{j\omega t} \\ \sum_{n,m,\omega} a_{0n} \frac{1}{r}\frac{\partial}{\partial r}\left(\sqrt{kr}H^{(2)}_{n+\frac{1}{2}}(kr)\right)\frac{\partial}{\partial\theta}P_n(\cos\theta) \cdot e^{j\omega t} \\ 0 \end{pmatrix}$$

$$\mathbf{H} = \begin{pmatrix} H_r \\ H_\theta \\ H_\varphi \end{pmatrix} = \begin{pmatrix} 0 \\ 0 \\ \sum_{n,m,\omega} a_{0n} \frac{\varepsilon j\omega}{r}\sqrt{kr}H^{(2)}_{n+\frac{1}{2}}(kr) P^1_n(\cos\theta) \cdot e^{j\omega t} \end{pmatrix}$$

Interesting to consider $n = 1$ where $H^{(2)}_{\frac{3}{2}}(kr) = \sqrt{\frac{2}{\pi kr}}\,\mathrm{e}^{-jkr}(\frac{j}{kr} - 1)$.

$$\mathbf{E} = \begin{pmatrix} E_r \\ E_\theta \\ E_\varphi \end{pmatrix} = \begin{pmatrix} a_{01} jk^2 2\sqrt{\frac{2}{\pi}}\,\mathrm{e}^{-jkr}\left(\frac{j}{(kr)^2} + \frac{1}{(kr)^3}\right)\cos\theta \cdot e^{j\omega t} \\ a_{01} jk^2 \sqrt{\frac{2}{\pi}}\,\mathrm{e}^{-jkr}\left(-\frac{1}{kr} + \frac{j}{(kr)^2} + \frac{1}{(kr)^3}\right)\sin\theta \cdot e^{j\omega t} \\ 0 \end{pmatrix}$$

$$\mathbf{H} = \begin{pmatrix} H_r \\ H_\theta \\ H_\varphi \end{pmatrix} = \begin{pmatrix} 0 \\ 0 \\ a_{01}\varepsilon j\omega k\sqrt{\frac{2}{\pi}}\,\mathrm{e}^{-jkr}\left(-\frac{1}{kr} + \frac{j}{(kr)^2}\right)\sin\theta \cdot e^{j\omega t} \end{pmatrix}$$

Except for the constants, the field is identical to that of an elementary dipole antenna. Both cases show r raised to the same powers in their corresponding terms of the denominators.
Taking the ratio between the electric/magnetic amplitudes yields a result consistent to that for the elementary dipole antenna

$$\frac{a_{01} jk^2\sqrt{\frac{2}{\pi}}}{a_{01}\varepsilon j\omega k\sqrt{\frac{2}{\pi}}} = \frac{k}{\varepsilon\omega} = \frac{\omega\sqrt{\varepsilon\mu}}{\varepsilon\omega} = \sqrt{\frac{\mu}{\varepsilon}}$$

It may be worth writing

$$\mathbf{E} = \begin{pmatrix} E_r \\ E_\theta \\ E_\varphi \end{pmatrix} = Z_0 H_0 \begin{pmatrix} \left(\frac{2j}{(kr)^2} + \frac{2}{(kr)^3}\right)\cos\theta \cdot e^{j(\omega t - kr)} \\ \left(-\frac{1}{(kr)} + \frac{j}{(kr)^2} + \frac{1}{(kr)^3}\right)\sin\theta \cdot e^{j(\omega t - kr)} \\ 0 \end{pmatrix} \qquad Z_0 = \sqrt{\frac{\mu_0}{\varepsilon_0}} \approx 120\pi\Omega \text{ field resistance}$$

$$\mathbf{H} = \begin{pmatrix} H_r \\ H_\theta \\ H_\varphi \end{pmatrix} = H_0 \begin{pmatrix} 0 \\ 0 \\ \left(-\frac{1}{(kr)} + \frac{j}{(kr)^2}\right)\sin\theta \cdot e^{j(\omega t - kr)} \end{pmatrix} \qquad H_0 = \begin{cases} \frac{I_0 l}{4\pi} k^2 \frac{1}{j} & l << \lambda \\ a_{01}\varepsilon_0 j\omega k\sqrt{\frac{2}{\pi}} & m = 0, n = 1, TM \end{cases}$$

For the farfield, i.e. ignoring 2nd and 3rd order terms, it may be written $\mathbf{E} = Z_0\ \mathbf{H} \times \mathbf{n}$ where $\mathbf{n}$ is the normal vector (pointing in, i.e. aligning with, the direction of propagation).

Letting $a_{01}\varepsilon_0 j\omega k\sqrt{\frac{2}{\pi}} = \frac{I_0 l}{4\pi} k^2 \frac{1}{j}$ would write $a_{01} = \frac{1}{j\sqrt{\frac{2}{\pi}}}\frac{I_0 l}{4\pi}\frac{k}{\varepsilon_0\omega}\frac{1}{j} = \frac{1}{j\sqrt{\frac{2}{\pi}}}\frac{I_0 l}{4\pi}\frac{\omega\sqrt{\varepsilon_0\mu_0}}{\varepsilon_0\omega}\frac{1}{j}$ to read

$a_{01} = \frac{1}{j\sqrt{\frac{2}{\pi}}}\frac{I_0 l}{4\pi}\sqrt{\frac{\mu_0}{\varepsilon_0}}\frac{1}{j}$ in terms of the dipole's params.

§§§

Cross Relationships

Constructing cylindrical waves from plane waves

Given the plane wave, Cf. 1.)

$$\text{RHS} \equiv \mathrm{e}^{j\omega t} \int_{\theta'} \int_{\varphi'} g(\theta',\varphi')\, \mathrm{e}^{jk\mathbf{nR}}\, d\varphi' d\theta'$$

RHS

$$\equiv \mathrm{e}^{j\omega t} \int_{\theta'} \int_{\varphi'} g(\theta',\varphi')\, \mathrm{e}^{jk(x\sin\theta'\cos\varphi' + y\sin\theta'\sin\varphi' + z\cos\theta')}\, d\varphi' d\theta'$$

For the expression shown it would make sense to express x, y, z of the RHS in terms of cylindrical coordinates r, φ, z, as is chosen to perform below, OR to express r, φ, z of the RHS in terms of Cartesian coordinates x, y, z.

It would make little sense to try to construct a cylindrical wave by rotating the planes around the x- or y-axis. However rotating the planes around the z-axis means that the z-component of $\mathbf{k}$ is a constant, $kn_z = k\cos\theta' = -\beta$. The plane waves from which to construct the cylindrical wave thus lie on a cone defined by a constant cone angle $2\theta'$ with its tip at the origin. The constant angle θ' means that φ' is the only variable left, over which to integrate. The plane wave becomes

$$Z(r,\varphi;t)$$

$$= \mathrm{e}^{j\omega t} \int_{\varphi'} g(\varphi')\, \mathrm{e}^{jk(x\sin\theta'\cos\varphi' + y\sin\theta'\sin\varphi')}\, d\varphi'$$

allowing an Ansatz to be approached at this stage as

$$Z(r,\varphi,z;t) = Z(r,\varphi)\, \mathrm{e}^{j(\omega t - \beta z)}$$

where

$$Z(r,\varphi) = \int_{\varphi'} g(\varphi')\, \mathrm{e}^{jk(x\sin\theta'\cos\varphi' + y\sin\theta'\sin\varphi')}\, d\varphi'$$

The conjecture is that $Z(r,\varphi)$ will be separable into (stand-alone) r and φ dependencies.

Expressing the RHS in terms of cylindrical coordinates r, φ using the substitutions

$$x = r\cos\varphi \qquad\qquad y = r\sin\varphi$$

the exponent is written

$$jk(x\sin\theta'\cos\varphi' + y\sin\theta'\sin\varphi')$$
$$= jk(r\cos\varphi\sin\theta'\cos\varphi' + r\sin\varphi\sin\theta'\sin\varphi')$$
$$= jkr\sin\theta'(\cos\varphi\cos\varphi' + \sin\varphi\sin\varphi')$$
$$= jkr\sin\theta'\cos(\varphi' - \varphi)$$

From $kn_z = k\cos\theta' = -\beta$ writing $\cos\theta' = \frac{-\beta}{k}$ and returning to $\sin\theta'$ for the exponent

$$\sin\theta' = \sqrt{1-\cos^2\theta'} = \sqrt{1-(\tfrac{-\beta}{k})^2} = \tfrac{\sqrt{k^2-\beta^2}}{k}$$

and thus

$$kr\sin\theta' = kr\tfrac{\sqrt{k^2-\beta^2}}{k} = r\sqrt{k^2-\beta^2} = \rho$$

The exponent becomes

$$jk(x\sin\theta'\cos\varphi' + y\sin\theta'\sin\varphi') = j\rho\cos(\varphi'-\varphi)\ Z(r,\varphi) = \int_{\varphi'} g(\varphi')\mathrm{e}^{j\rho\cos(\varphi'-\varphi)}\,d\varphi'$$

Introducing a new variable $\delta = \varphi' - \varphi$

$$Z(r,\varphi) = \int_{\delta} g(\delta+\varphi)\mathrm{e}^{j\rho\cos\delta}\,d\delta$$

The reasoning involves writing $g(\delta+\varphi) = g_1(\delta)g_2(\varphi)$
thus separating $Z(r,\varphi)$ into the following dependencies

$$Z(r,\varphi) = \Phi(\varphi)R(r) = g_2(\varphi)\int_{\delta} g_1(\delta)\mathrm{e}^{j\rho\cos\delta}\,d\delta$$

It follows that

$\Phi(\varphi) = g_2(\varphi) = \mathrm{e}^{\pm jn\varphi}$

also

$$R(r) = \int_{\delta} g_1(\delta)\mathrm{e}^{j\rho\cos\delta}\,d\delta$$

As an Ansatz, the following relationship is desirable,

$Z(r,\varphi,z;t) = R(r)\mathrm{e}^{jn\varphi}\mathrm{e}^{j(\omega t-\beta z)}$

where $R(r)$ would satisfy the following condition, the condition recognized as (a common version of) the Bessel differential equation

$$\frac{d^2R}{dr^2} + \frac{1}{r}\frac{dR}{dr} + \left((k^2-\beta^2) - \frac{n^2}{r^2}\right)R = 0$$

Substituting $\rho = r\sqrt{k^2-\beta^2}$ to get back to the normal form, Cf. 2.)

$$\frac{d^2R}{d\rho^2} + \frac{1}{\rho}\frac{dR}{d\rho} + \left(1 - \frac{n^2}{\rho^2}\right)R = 0$$

Performing the operations specified by the differential equation leads to, Cf. 3.),

$\int_{\delta} \left(\rho^2\sin^2\delta + j\rho\cos\delta - n^2\right)g_1(\delta)\mathrm{e}^{j\rho\cos\delta}\,d\delta = 0$ It may be verified, Cf. 4.), that the Eq. shown is equivalent to the following

$$\int_{\delta} \frac{d}{d\delta}\left(g_1(\delta)\frac{d\,\mathrm{e}^{j\rho\cos\delta}}{d\delta} - \frac{dg_1(\delta)}{d\delta}\mathrm{e}^{j\rho\cos\delta}\right)d\delta$$
$$+\int_{\delta}\left(\frac{d^2g_1(\delta)}{d\delta^2} + n^2g_1(\delta)\right)\mathrm{e}^{j\rho\cos\delta}\,d\delta = 0$$

$$\int_{\delta} \frac{d}{d\delta}\left(g_1(\delta)(-j\rho\sin\delta)\mathrm{e}^{j\rho\cos\delta} - \frac{dg_1(\delta)}{d\delta}\mathrm{e}^{j\rho\cos\delta}\right)d\delta \qquad +\int_{\delta}\left(\frac{d^2g_1(\delta)}{d\delta^2} + n^2g_1(\delta)\right)\mathrm{e}^{j\rho\cos\delta}\,d\delta = 0$$

$$\int_{\delta} \frac{d}{d\delta}\left([-j\rho\sin\delta\,g_1(\delta) - \frac{dg_1(\delta)}{d\delta}]\mathrm{e}^{j\rho\cos\delta}\right)d\delta$$
$$+\int_{\delta}\left(\frac{d^2g_1(\delta)}{d\delta^2} + n^2g_1(\delta)\right)\mathrm{e}^{j\rho\cos\delta}\,d\delta = 0$$

The 1st integral will vanish when the integration path is chosen such that the values for $[-j\rho\sin\delta\,g_1(\delta) - \frac{dg_1(\delta)}{d\delta}]\mathrm{e}^{j\rho\cos\delta}$ at the start and stop points are the same.

The 2nd integral will vanish if $g_1(\delta)$ satisfies the differential equation

$$\frac{d^2 g_1(\delta)}{d\delta^2} + n^2 g_1(\delta) = 0$$

the solution for which may be given as

$$g_1(\delta) = A\,e^{jn\delta}$$

On substituting

$$R(r) = \int_\delta A\,e^{jn\delta} e^{j\rho\cos\delta}\, d\delta$$

$$R(r) = A\int_\delta e^{j(\rho\cos\delta + n\delta)}\, d\delta$$

As it satisfies the Bessel differential equation, it is guessed that this is a line integral representation for the cylinder function (Wolfgang Gröbner & Nikolaus Hofreiter, Integraltafel, Bestimmte Integrale, Springer Verlag Wien Österreich,1950, 1958, 1961, 1966, revision of 1973, page 189, Eq.12b)

$$Z_n(\rho) = \frac{e^{-jn\frac{\pi}{2}}}{2\pi}\int_L e^{j(\rho\cos\delta + n\delta)}\, d\delta$$

whereby using $g_1(\delta) = A\,e^{jn\delta}$ in $[-j\rho\sin\delta\, g_1(\delta) - \frac{dg_1(\delta)}{d\delta}]e^{j\rho\cos\delta}$ and requiring $[\rho\sin\delta + n]e^{j(\rho\cos\delta + n\delta)}$ to have the same values at the start and stop points of L

For integer n, L will be between $-\pi$ and $+\pi$ on the real axis and the requirement is readily satisfied.

The cylinder function becomes

$$J_n(\rho) = \frac{e^{-jn\frac{\pi}{2}}}{2\pi}\int_{-\pi}^{+\pi} e^{j(\rho\cos\delta + n\delta)}\, d\delta$$

$$J_n(\rho) = \frac{j^{-n}}{2\pi}\int_{-\pi}^{+\pi} e^{j(\rho\cos\delta + n\delta)}\, d\delta$$

Constructing plane waves from cylindrical waves

Converting from spherical and cylindrical respectively to express $\mathbf{k}$ and $\mathbf{R}$ in Cartesian

$$\mathbf{k} = \begin{pmatrix} k_x \\ k_y \\ k_z \end{pmatrix} = k\mathbf{n} = k\begin{pmatrix} \sin\theta'\cos\varphi' \\ \sin\theta'\sin\varphi' \\ \cos\theta' \end{pmatrix}, \mathbf{R} = \begin{pmatrix} r\cos\varphi \\ r\sin\varphi \\ z \end{pmatrix}$$

Ansatz for the plane wave: $e^{j(\omega t - k[\mathbf{nR}])}$

$\mathbf{nR} =$

$r\cos\varphi\sin\theta'\cos\varphi' + r\sin\varphi\sin\theta'\sin\varphi' + z\cos\theta'$

$= r\sin\theta'(\cos\varphi\cos\varphi' + \sin\varphi\sin\varphi') + z\cos\theta'$

$= r\sin\theta'\cos(\varphi' - \varphi) + z\cos\theta'$

Thus

$e^{j(\omega t - k[\mathbf{nR}])} = e^{j(\omega t - k[r\sin\theta'\cos(\varphi'-\varphi) + z\cos\theta'])}$

$e^{j(\omega t - k[\mathbf{nR}])} = e^{-jkr\sin\theta'\cos(\varphi'-\varphi)} e^{j(\omega t - kz\cos\theta')}$

$e^{j(\omega t - k[\mathbf{nR}])} = f(r,\varphi)\, e^{j(\omega t - kz\cos\theta')}$

where $f(r,\varphi) = e^{-jkr\sin\theta'\cos(\varphi'-\varphi)}$

Ansatz: Expanding $f(r,\varphi)$ into a Fourier series with Fourier coefficients $f_n(r)$, Cf. 5.)

$$f(r,\varphi)=\sum_{-\infty}^{+\infty} f_n(r)\,\mathrm{e}^{jn\varphi} \qquad \text{Inverse}$$

$$f_n(r)=\frac{1}{2\pi}\int_{-\pi}^{+\pi} f(r,\varphi)\,\mathrm{e}^{-jn\varphi}\,d\varphi \qquad \text{Direct}$$

$$f_n(r)=\frac{1}{2\pi}\int_{-\pi}^{+\pi} \mathrm{e}^{-jkr\sin\theta'\cos(\varphi'-\varphi)}\mathrm{e}^{-jn\varphi}\,d\varphi$$

$$f_n(r)=\frac{1}{2\pi}\int_{-\pi}^{+\pi} \mathrm{e}^{j[-kr\sin\theta'\cos(\varphi'-\varphi)-n\varphi]}\,d\varphi$$

Substituting $\varphi'-\varphi=\delta$, $\varphi=\varphi'-\delta$, $d\varphi=-d\delta$

$$f_n(r)=\frac{1}{2\pi}\int_{\varphi'+\pi}^{\varphi'-\pi} \mathrm{e}^{j[-kr\sin\theta'\cos\delta-n(\varphi'-\delta)]}(-d\delta)$$

$$f_n(r)=\frac{1}{2\pi}\int_{\varphi'-\pi}^{\varphi'+\pi} \mathrm{e}^{j[-kr\sin\theta'\cos\delta-n(\varphi'-\delta)]}\,d\delta$$

$$f_n(r)=\frac{1}{2\pi}\int_{-\pi}^{+\pi} \mathrm{e}^{j[-kr\sin\theta'\cos\delta-n(\varphi'-\delta)]}\,d\delta$$

Bringing it to the form known for $J_n(\rho)$ above

$$f_n(r)=\mathrm{e}^{-jn\varphi'}\frac{1}{2\pi}\int_{-\pi}^{+\pi} \mathrm{e}^{j[-kr\sin\theta'\cos\delta+n\delta]}\,d\delta$$

Multiplying the RHS by $1=j^{+n}j^{-n}$

$$f_n(r)=\mathrm{e}^{-jn\varphi'}\frac{1}{2\pi}j^{+n}j^{-n}\int_{-\pi}^{+\pi} \mathrm{e}^{j[-kr\sin\theta'\cos\delta+n\delta]}\,d\delta$$

$$f_n(r)=\mathrm{e}^{-jn\varphi'}j^{+n}\frac{j^{-n}}{2\pi}\int_{-\pi}^{+\pi} \mathrm{e}^{j[-kr\sin\theta'\cos\delta+n\delta]}\,d\delta$$

Using $j^{+n}=\mathrm{e}^{jn\frac{\pi}{2}}$ and the expression for $J_n(\rho)$ above

$$f_n(r)=\mathrm{e}^{-jn\varphi'}j^{+n}J_n(-kr\sin\theta')$$

$$f_n(r)=\mathrm{e}^{-jn\varphi'}\mathrm{e}^{jn\frac{\pi}{2}}J_n(-kr\sin\theta')$$

$$f_n(r)=\mathrm{e}^{jn\left(\frac{\pi}{2}-\varphi'\right)}J_n(-kr\sin\theta')$$

furthermore

$$f_n(r)=\mathrm{e}^{jn\left(\frac{\pi}{2}-\varphi'\right)}(-1)^n J_n(kr\sin\theta')$$

Recall that

$$\mathrm{e}^{j(\omega t-k[\mathbf{nR}])}=f(r,\varphi)\,\mathrm{e}^{j(\omega t-kz\cos\theta')}$$

where $f(r,\varphi)=\mathrm{e}^{-jkr\sin\theta'\cos(\varphi'-\varphi)}$

$$f(r,\varphi)=\sum_{-\infty}^{+\infty} f_n(r)\,\mathrm{e}^{jn\varphi}$$

thus

$$f(r,\varphi)=\sum_{-\infty}^{+\infty} \mathrm{e}^{jn\frac{\pi}{2}}(-1)^n J_n(kr\sin\theta')\,\mathrm{e}^{jn(\varphi-\varphi')}$$

$$j^{+n} = e^{jn\frac{\pi}{2}}$$

$$f(r,\varphi) = \sum_{-\infty}^{+\infty} j^{+n}(-1)^n J_n(kr\sin\theta')\,e^{jn(\varphi-\varphi')}$$

$$f(r,\varphi) = \sum_{-\infty}^{+\infty} (-j)^n J_n(kr\sin\theta')\,e^{jn(\varphi-\varphi')}$$

From $e^{j(\omega t-k[\mathbf{nR}])} = f(r,\varphi)\,e^{j(\omega t-kz\cos\theta')}$, a plane wave has been constructed from cylindrical waves by superposition, in particular

$$e^{j(\omega t-k[\mathbf{nR}])} =$$

$$\sum_{-\infty}^{+\infty} (-j)^n J_n(kr\sin\theta')\,e^{jn(\varphi-\varphi')}e^{j(\omega t-kz\cos\theta')}$$

For simplicity the same may be written

$$e^{-jk\mathbf{nR}} = \sum_{-\infty}^{+\infty} (-j)^n J_n(kr\sin\theta')\,e^{jn(\varphi-\varphi')}e^{-jkz\cos\theta'}$$

For the following special case, recalling that $f(r,\varphi) = e^{-jkr\sin\theta'\cos(\varphi'-\varphi)}$, so

$$f(r,\varphi) =$$

$$e^{-jkr\sin\theta'\cos(\varphi'-\varphi)} = \sum_{-\infty}^{+\infty} (-j)^n J_n(kr\sin\theta')\,e^{jn(\varphi-\varphi')}$$

Letting $\rho = -kr\sin\theta'$ and $\varphi-\varphi' = \alpha - \frac{\pi}{2}$

$$e^{-jkr\sin\theta'\cos(\alpha-\frac{\pi}{2})} = \sum_{-\infty}^{+\infty} (-j)^n J_n(-\rho)\,e^{jn(\alpha-\frac{\pi}{2})}$$

$$e^{j\rho\sin\alpha} = \sum_{-\infty}^{+\infty} (-j)^n J_n(-\rho)\,e^{jn\alpha}e^{-jn\frac{\pi}{2}}$$

$$e^{j\rho\sin\alpha} = \sum_{-\infty}^{+\infty} (-j)^n(-1)^n J_n(\rho)\,e^{jn\alpha}\frac{1}{j^n}$$

$$e^{j\rho\sin\alpha} = \sum_{-\infty}^{+\infty} j^n J_n(\rho)\,e^{jn\alpha}\frac{1}{j^n}$$

$$e^{j\rho\sin\alpha} = \sum_{-\infty}^{+\infty} J_n(\rho)\,e^{jn\alpha}$$

Separating the real-parts from the imaginary parts of both sides

$$\cos(\rho\sin\alpha) = \sum_{-\infty}^{+\infty} J_n(\rho)\cos(n\alpha)$$

$$\sin(\rho\sin\alpha) = \sum_{-\infty}^{+\infty} J_n(\rho)\sin(n\alpha)$$

Constructing plane waves from spherical waves

Starting out from the following, Cf. 6.) where $\cos\gamma = \cos\theta$ is assumed so as to first consider propagation in the z-axis direction,

$$e^{-jkR\cos\gamma} =$$

$$\sum_{n=0}^{\infty} (-j)^n(2n+1)\sqrt{\frac{\frac{\pi}{2}}{kR}}\,J_{n+\frac{1}{2}}(kR)P_n(\cos\theta)$$

that represents a series expansion for a plane wave that propagates in the z-axis direction

By the following theorem, $P_n(\cos\theta)$ may be expressed further as shown

$$P_n(\cos\gamma) = P_n(\cos\theta)P_n(\cos\theta') +$$

$$+2\sum_{m=1}^{n} \frac{(n-m)!}{(n+m)!} P_n^m(\cos\theta)P_n^m(\cos\theta')\cos m(\varphi-\varphi')$$

thus describing propagation in any arbitrary direction

Assembling the expression

$$e^{-jkR\cos\gamma} =$$

$$\sum_{n=0}^{\infty} (-j)^n (2n+1)\sqrt{\frac{\frac{\pi}{2}}{kR}} J_{n+\frac{1}{2}}(kR)\Big[P_n(\cos\theta)P_n(\cos\theta')$$

$$+2\sum_{m=1}^{n} \tfrac{(n-m)!}{(n+m)!} P_n^m(\cos\theta)P_n^m(\cos\theta')\cos m(\varphi-\varphi')\Big]$$

Constructing spherical waves from plane waves

Try again from

$$e^{-jkR\cos\gamma} =$$

$$\sum_{n=0}^{\infty} (-j)^n (2n+1)\sqrt{\frac{\frac{\pi}{2}}{kR}} J_{n+\frac{1}{2}}(kR) P_n(\cos\theta)$$

Multiplying both sides with $P_m(\cos\theta)$ and integrating over the surface of the unit sphere

$$\int_{\varphi=0}^{2\pi}\int_{\theta=0}^{\pi} e^{-jkR\cos\gamma}\, P_m(\cos\theta)\sin\theta d\theta d\varphi =$$

$$\int_{\varphi=0}^{2\pi}\int_{\theta=0}^{\pi} \sum_{n=0}^{\infty} (-j)^n (2n+1)\sqrt{\frac{\frac{\pi}{2}}{kR}} J_{n+\frac{1}{2}}(kR) P_n(\cos\theta) P_m(\cos\theta)\sin\theta d\theta d\varphi$$

thus

$$2\pi\int_{\theta=0}^{\pi} e^{-jkR\cos\gamma}\, P_m(\cos\theta)\sin\theta d\theta =$$

$$2\pi\sum_{n=0}^{\infty} (-j)^n (2n+1)\sqrt{\frac{\frac{\pi}{2}}{kR}} J_{n+\frac{1}{2}}(kR)\int_{\theta=0}^{\pi} P_n(\cos\theta)P_m(\cos\theta)\sin\theta d\theta$$

For the RHS consider

$$\int_{\theta=0}^{\pi} P_n(\cos\theta)P_m(\cos\theta)\sin\theta d\theta$$

where letting $x=\cos\theta$, $dx=-\sin\theta d\theta$

$$\int_{\theta=0}^{\pi} P_n(\cos\theta)P_m(\cos\theta)\sin\theta d\theta = \int_{x=1}^{x=-1} P_n(x)P_m(x)(-dx)$$

$$= \int_{-1}^{1} P_n(x)P_m(x)dx = \begin{cases} 0 & n\neq m \\ \frac{2}{2m+1} & n=m \end{cases}$$

giving

$$2\pi\int_{\theta=0}^{\pi} e^{-jkR\cos\gamma}\, P_m(\cos\theta)\sin\theta d\theta =$$

$$2\pi(-j)^m (2m+1)\sqrt{\frac{\frac{\pi}{2}}{kR}} J_{m+\frac{1}{2}}(kR)\frac{2}{2m+1}$$

on cleaning up

$$\sqrt{\frac{\frac{\pi}{2}}{kR}}J_{m+\frac{1}{2}}(kR) =$$

$$\frac{1}{2(-j)^m}\int_{\theta=0}^{\pi} e^{-jkR\cos\theta}\, P_m(\cos\theta)\sin\theta d\theta$$

Recapture

Cylindrical $\Leftarrow$ Plane waves

$$R(r)e^{jn\varphi}e^{j(\omega t-\beta z)}$$

$$= e^{j\omega t}\int_{\theta'}\int_{\varphi'} g(\theta',\varphi')e^{jk(x\sin\theta'\cos\varphi'+y\sin\theta'\sin\varphi'+z\cos\theta')}\, d\varphi' d\theta' \quad J_n(\rho) = \frac{j^{-n}}{2\pi}\int_{-\pi}^{+\pi} e^{j(\rho\cos\delta+n\delta)}\, d\delta$$

Plane waves $\Leftarrow$ Cylindrical

$$e^{-jk\mathbf{nR}} = \sum_{-\infty}^{+\infty} (-j)^n J_n(kr\sin\theta')e^{jn(\varphi-\varphi')}e^{-jkz\cos\theta'}$$

Plane waves $\Leftarrow$ Spherical

$$e^{-jkR\cos\gamma} =$$

$$\sum_{n=0}^{\infty} (-j)^n (2n+1)\sqrt{\frac{\frac{\pi}{2}}{kR}}J_{n+\frac{1}{2}}(kR)\Big[P_n(\cos\theta)P_n(\cos\theta')$$

$$+2\sum_{m=1}^{n} \frac{(n-m)!}{(n+m)!} P_n^m(\cos\theta)P_n^m(\cos\theta')\cos m(\varphi-\varphi')\Big]$$

Spherical $\Leftarrow$ Plane waves

$$\sqrt{\frac{\frac{\pi}{2}}{kR}}J_{m+\frac{1}{2}}(kR) =$$

$$\frac{1}{2(-j)^m}\int_{\theta=0}^{\pi} e^{-jkR\cos\theta}\, P_m(\cos\theta)\sin\theta d\theta$$

Addendum

1.) Variable usage:

$\mathbf{k}$ or the unit vector $\mathbf{n}$, the direction of propagation, also coinciding with the direction of point Q drawn from the origin, the latter would usually be expressed in spherical coordinates r', θ', φ', the Cartesian of which may thus be given

$$\mathbf{k} = \begin{pmatrix} k_x \\ k_y \\ k_z \end{pmatrix} = k \begin{pmatrix} n_x \\ n_y \\ n_z \end{pmatrix}, \quad \mathbf{n} = \begin{pmatrix} n_x \\ n_y \\ n_z \end{pmatrix} = \begin{pmatrix} \sin\theta' \cos\varphi' \\ \sin\theta' \sin\varphi' \\ \cos\theta' \end{pmatrix}$$

$k^2 = -(\sigma + j\omega\varepsilon) j\omega\mu \equiv k_x^2 + k_y^2 + k_z^2$ from $k = \sqrt{-(\sigma + j\omega\varepsilon) j\omega\mu}$

Point P, the p.o.i, likewise Q, may be written

$$\mathbf{P} = \begin{pmatrix} x \\ y \\ z \end{pmatrix} \qquad \mathbf{Q} = \begin{pmatrix} \xi \\ \eta \\ \zeta \end{pmatrix} \qquad \mathbf{r} = \mathbf{Q} - \mathbf{P}$$

γ is the angle between $\mathbf{P}$ and $\mathbf{Q}$

An Ansatz for plane waves may be made by superposition, for example

$$\psi(x,y,z;t) = \mathrm{e}^{j\omega t} \int_{\theta'} \int_{\varphi'} g(\theta',\varphi') \mathrm{e}^{jk\mathbf{nR}} \, d\varphi' d\theta'$$

$$= \mathrm{e}^{j\omega t} \int_{\theta'} \int_{\varphi'} g(\theta',\varphi') \mathrm{e}^{jk(n_x x + n_y y + n_z z)} \, d\varphi' d\theta'$$

$$= \mathrm{e}^{j\omega t} \int_{\theta'} \int_{\varphi'} g(\theta',\varphi') \mathrm{e}^{jk(x\sin\theta'\cos\varphi' + y\sin\theta'\sin\varphi' + z\cos\theta')} \, d\varphi' d\theta'$$

The use of variable or function names is not as strictly consistent as would probably be expected, e.g. $\psi(x,y,z;t)$ is written interchangeably as $\psi(r,\varphi,z;t)$ the latter would involve cylindrical instead of Cartesian coordinates. Also, for simplicity, any vars not being used would be dropped from the list of arguments.

2.) Checking variable substitution for Bessel differential equation:

$d\rho = \sqrt{k^2 - \beta^2}\, dr$, $d\rho^2 = (k^2 - \beta^2) dr^2$

$$\frac{d^2R}{(k^2 - \beta^2)dr^2} + \frac{1}{r\sqrt{k^2 - \beta^2}} \frac{dR}{\sqrt{k^2 - \beta^2}\, dr}$$

$$+ \left(1 - \frac{n^2}{r^2(k^2 - \beta^2)}\right) R = 0$$

$$\frac{d^2R}{(k^2 - \beta^2)dr^2} + \frac{1}{r(k^2 - \beta^2)} \frac{dR}{dr}$$

$$+ \left(1 - \frac{n^2}{r^2(k^2 - \beta^2)}\right) R = 0$$

$$\frac{d^2R}{dr^2} + \frac{1}{r}\frac{dR}{dr} + \left((k^2 - \beta^2) - \frac{n^2}{r^2}\right) R = 0$$

checked OK.

3.) Carrying out the operations for

$$\frac{d^2R}{d\rho^2}+\frac{1}{\rho}\frac{dR}{d\rho}+\left(1-\frac{n^2}{\rho^2}\right)R=0$$

$$\frac{dR}{d\rho}=\int \frac{d}{d\rho}g_1(\delta)\,\mathrm{e}^{j\rho\cos\delta}\,d\delta$$

$$=\int g_1(\delta)\,j\cos\delta\,\mathrm{e}^{j\rho\cos\delta}\,d\delta$$

$$\frac{d^2R}{d\rho^2}=\int g_1(\delta)(-\cos^2\delta)\,\mathrm{e}^{j\rho\cos\delta}\,d\delta$$

Assembling the expression

$$\int g_1(\delta)(-\cos^2\delta)\,\mathrm{e}^{j\rho\cos\delta}\,d\delta+$$

$$+\frac{1}{\rho}\int g_1(\delta)\,j\cos\delta\,\mathrm{e}^{j\rho\cos\delta}\,d\delta$$

$$+(1-\frac{n^2}{\rho^2})\int g_1(\delta)\,\mathrm{e}^{j\rho\cos\delta}\,d\delta=0$$

$$\int g_1(\delta)[(-\cos^2\delta)+$$

$$+\frac{1}{\rho}j\cos\delta$$

$$+(1-\frac{n^2}{\rho^2})]\,\mathrm{e}^{j\rho\cos\delta}\,d\delta=0$$

Multiplying both sides with ρ^2

$$\int g_1(\delta)[(-\cos^2\delta)+\frac{1}{\rho}j\cos\delta$$

$$+(1-\frac{n^2}{\rho^2})]\,\mathrm{e}^{j\rho\cos\delta}\,d\delta=0$$

$$\int g_1(\delta)[(-\rho^2\cos^2\delta)+\rho j\cos\delta$$

$$+(\rho^2-n^2)]\,\mathrm{e}^{j\rho\cos\delta}\,d\delta=0$$

$\int g_1(\delta)[\rho^2\sin^2\delta+j\rho\cos\delta-n^2]\,\mathrm{e}^{j\rho\cos\delta}\,d\delta=0$ checked OK.

4.) Checking expression

$$\int_\delta \frac{d}{d\delta}\left(g_1(\delta)\frac{d\,\mathrm{e}^{j\rho\cos\delta}}{d\delta}-\frac{dg_1(\delta)}{d\delta}\mathrm{e}^{j\rho\cos\delta}\right)d\delta$$

$$+\int_\delta\left(\frac{d^2g_1(\delta)}{d\delta^2}+n^2g_1(\delta)\right)\mathrm{e}^{j\rho\cos\delta}\,d\delta=0$$

Expression under the 1st integral:

$$g_1(\delta)\frac{d\,\mathrm{e}^{j\rho\cos\delta}}{d\delta}-\frac{dg_1(\delta)}{d\delta}\mathrm{e}^{j\rho\cos\delta}=$$

$$=\frac{d}{d\delta}g_1\frac{d}{d\delta}\mathrm{e}^{j\rho\cos\delta}+g_1\frac{d^2}{d\delta^2}\mathrm{e}^{j\rho\cos\delta}$$

$$-\mathrm{e}^{j\rho\cos\delta}\frac{d^2}{d\delta^2}g_1-\frac{d}{d\delta}\mathrm{e}^{j\rho\cos\delta}\frac{d}{d\delta}g_1$$

$$=g_1\frac{d^2}{d\delta^2}\mathrm{e}^{j\rho\cos\delta}-\mathrm{e}^{j\rho\cos\delta}\frac{d^2}{d\delta^2}g_1$$

$$\frac{d^2}{d\delta^2}\mathrm{e}^{j\rho\cos\delta}=\frac{d}{d\delta}(-j\rho\sin\delta\,\mathrm{e}^{j\rho\cos\delta})$$

$$=-j\rho\cos\delta\,\mathrm{e}^{j\rho\cos\delta}-\rho^2\sin^2\delta\,\mathrm{e}^{j\rho\cos\delta}$$

$$g_1\frac{d^2}{d\delta^2}\mathrm{e}^{j\rho\cos\delta}=$$

$$=-j\rho\cos\delta g_1\,\mathrm{e}^{j\rho\cos\delta}-\rho^2\sin^2\delta g_1\,\mathrm{e}^{j\rho\cos\delta}$$

Complete expression under the 1st integral:

$$\frac{d}{d\delta}\left(g_1(\delta)\frac{d\,\mathrm{e}^{j\rho\cos\delta}}{d\delta}-\frac{dg_1(\delta)}{d\delta}\mathrm{e}^{j\rho\cos\delta}\right)=$$

$$=-j\rho\cos\delta g_1\,\mathrm{e}^{j\rho\cos\delta}-\rho^2\sin^2\delta g_1\,\mathrm{e}^{j\rho\cos\delta}$$

$$-\mathrm{e}^{j\rho\cos\delta}\frac{d^2}{d\delta^2}g_1$$

Expression under the 2nd integral:

$$\left(\frac{d^2g_1(\delta)}{d\delta^2}+n^2g_1(\delta)\right)\mathrm{e}^{j\rho\cos\delta}=$$

$$=\mathrm{e}^{j\rho\cos\delta}\frac{d^2}{d\delta^2}g_1+\mathrm{e}^{j\rho\cos\delta}n^2g_1$$

Writing both expressions under the integral integrating w.r.t δ

$$\int_\delta\left(\frac{d}{d\delta}[\cdots]+[\cdots]\right)d\delta=$$

checked OK.

$$=\int_\delta\left(-j\rho\cos\delta-\rho^2\sin^2\delta+n^2\right)g_1\,\mathrm{e}^{j\rho\cos\delta}\,d\delta=0$$

5.) Fourier integral expressions

$$F(f)=\int f(t)\mathrm{e}^{-j2\pi ft}\,dt$$

$$f(t)=\int F(f)\mathrm{e}^{+j2\pi ft}\,df$$

The expressions shown represent the pair known to electrical engineers

By the variable substitution $\omega=2\pi f$, $f=\frac{\omega}{2\pi}$, $df=\frac{1}{2\pi}d\omega$, the 2nd integral is written

$$f(t)=\int\frac{F(\frac{\omega}{2\pi})}{2\pi}\mathrm{e}^{+j\omega t}\,d\omega$$

Letting $G(\omega)\equiv\frac{F(\frac{\omega}{2\pi})}{2\pi}$, and dividing both sides of the 1st integral by 2π to write

$$\frac{F(\frac{\omega}{2\pi})}{2\pi} = \frac{1}{2\pi}\int f(t)\mathrm{e}^{-j\omega t}\,dt$$

and to arrive at the following equivalent pair

$$G(\omega) = \frac{1}{2\pi}\int f(t)\mathrm{e}^{-j\omega t}\,dt$$

$$f(t) = \int G(\omega)\mathrm{e}^{+j\omega t}\,d\omega$$

6.) Evaluating Legendre polynomials' integrals

On the z-axis $\theta' = 0$, $\cos\gamma = \cos\theta$, with R being the spherical radius, the plane wave may be given as $\mathrm{e}^{-jkR\cos\gamma}$

$$\mathrm{e}^{-jkR\cos\gamma} = \sum_{n=0}^{\infty} a_n \sqrt{\frac{\frac{\pi}{2}}{kR}} J_{n+\frac{1}{2}}(kR) P_n(\cos\theta)$$

Letting $\cos\gamma = \cos\theta \equiv x$ in the following to work out the coefficients a_n, utilizing the Legendre polynomials' orthonormality

$$\int_{-1}^{+1} \mathrm{e}^{-jkRx} P_m(x)dx$$

$$= \int_{-1}^{+1} \sum_{n=0}^{\infty} a_n \sqrt{\frac{\frac{\pi}{2}}{kR}} J_{n+\frac{1}{2}}(kR) P_n(x) P_m(x) dx$$

The LHS evaluates as (Wolfgang Gröbner & Nikolaus Hofreiter, Integraltafel, Bestimmte Integrale, Springer Verlag Wien Österreich, 1950, 1958, 1961, 1966, revision of 1973, page 24, Eq.10)

$$\int_{-1}^{+1} \mathrm{e}^{-jkRx} P_m(x)dx = (-j)^m \sqrt{\frac{2\pi}{kR}} J_{m+\frac{1}{2}}(kR)$$

$$= (-j)^m \sqrt{\frac{4\frac{\pi}{2}}{kR}} J_{m+\frac{1}{2}}(kR) = (-j)^m 2\sqrt{\frac{\frac{\pi}{2}}{kR}} J_{m+\frac{1}{2}}(kR)$$

The RHS evaluates as

$$\int_{-1}^{+1} \sum_{n=0}^{\infty} a_n \sqrt{\frac{\frac{\pi}{2}}{kR}} J_{n+\frac{1}{2}}(kR) P_n(x) P_m(x) dx \quad = \sum_{n=0}^{\infty} a_n \sqrt{\frac{\frac{\pi}{2}}{kR}} J_{n+\frac{1}{2}}(kR) \int_{-1}^{+1} P_n(x) P_m(x) dx$$

$$= a_m \sqrt{\frac{\frac{\pi}{2}}{kR}} J_{m+\frac{1}{2}}(kR) \int_{-1}^{+1} P_m(x) P_m(x) dx$$

$$= a_m \sqrt{\frac{\frac{\pi}{2}}{kR}} J_{m+\frac{1}{2}}(kR) \frac{2}{2m+1}$$

Assembling both sides

$$(-j)^m 2\sqrt{\frac{\frac{\pi}{2}}{kR}} J_{m+\frac{1}{2}}(kR)$$

$$= a_m \sqrt{\frac{\frac{\pi}{2}}{kR}} J_{m+\frac{1}{2}}(kR) \frac{2}{2m+1}$$

giving

$$a_m = (-j)^m (2m+1)$$

The series expansion is now reading

$e^{-jkR\cos\gamma} =$

$$\sum_{n=0}^{\infty} (-j)^n (2n+1)\sqrt{\frac{\frac{\pi}{2}}{kR}} J_{n+\frac{1}{2}}(kR) P_n(\cos\theta)$$

7.) More of it

As $\frac{1}{r^2}\frac{d}{dr}(r^2\frac{d\psi}{dr}) = \frac{1}{r}\frac{d^2}{dr^2}(r\psi)$

The differential equation for Π_r and that for a scalar ψ (in particular $div\ grad\ \psi + k^2\psi = 0$ in spherical coordinates) led to separable equations that respectively involved $\frac{r^2}{R}\frac{d^2}{dr^2}(R) + k^2r^2 - n(n+1) = 0$ and $\frac{r^2}{R_\psi}\frac{1}{r}\frac{d^2}{dr^2}(rR_\psi) + k^2r^2 - n(n+1) = 0$ that are identical equations if rR_ψ is substituted by another scalar

The solution for the former is known as

$R(r) = \sqrt{kr}Z_{n+\frac{1}{2}}(kr)$ giving

$\Pi_r(r,\theta,\varphi;t) = \sqrt{kr}Z_{n+\frac{1}{2}}(kr)P_n^m(\cos\theta)e^{\pm jm\varphi}$

Thus the solution for the latter may be given

$R_\psi(r) = \frac{1}{r}\sqrt{kr}Z_{n+\frac{1}{2}}(kr)$ giving

$\psi(r,\theta,\varphi;t) = \frac{1}{r}\sqrt{kr}Z_{n+\frac{1}{2}}(kr)P_n^m(\cos\theta)e^{\pm jm\varphi}$

By linearity the solution to ψ is however taken to be

$\psi(r,\theta,\varphi;t) = \sqrt{\frac{\pi}{2kr}}Z_{n+\frac{1}{2}}(kr)P_n^m(\cos\theta)e^{\pm jm\varphi}$

The discrepancy is taken into account by the coefficients when superposition comes into play.

8.) Bessel revisited

(Wolfgang Gröbner & Nikolaus Hofreiter, Integraltafel, Bestimmte Integrale, Springer Verlag Wien Österreich, 1950, 1958, 1961, 1966, revision of 1973, page 188, Eqs.8d 8e)

$$J_{\nu+n}(z) = (-1)^n z^{\nu+n}\left(\frac{d}{zdz}\right)^n \left(z^{-\nu}J_\nu(z)\right) \qquad n = 0,1,2,\cdots$$

$$J_{\nu-n}(z) = z^{n-\nu}\left(\frac{d}{zdz}\right)^n \left(z^{\nu}J_\nu(z)\right) \qquad n = 0,1,2,\cdots$$

Letting $n = 1$

$J_{\nu+1}(z) = -z^{\nu+1}\frac{d}{zdz}\left(z^{-\nu}J_\nu(z)\right) = -z^{\nu}\frac{d}{dz}\left(z^{-\nu}J_\nu(z)\right)$ i.e. $\frac{d}{dz}\left(z^{-\nu}J_\nu(z)\right) = -z^{-\nu}J_{\nu+1}(z)$

$J_{\nu-1}(z) = z^{1-\nu}\frac{d}{zdz}\left(z^{\nu}J_\nu(z)\right) = z^{-\nu}\frac{d}{dz}\left(z^{\nu}J_\nu(z)\right)$ i.e. $\frac{d}{dz}\left(z^{\nu}J_\nu(z)\right) = z^{\nu}J_{\nu-1}(z)$

Starting from

$\frac{d}{dx}(x^{-n}Z_n(x)) = -x^{-n}Z_{n+1}(x)$, Also $\frac{d}{dx}(x^nZ_n(x)) = x^nZ_{n-1}(x)$, $Z_n(x)$ may be written as Z_n for short

Consider the first one LHS $\frac{d}{dx}(x^{-n}Z_n(x)) = -nx^{-n-1}Z_n + x^{-n}Z_n'$ equating to RHS $-nx^{-n-1}Z_n + x^{-n}Z_n' = -x^{-n}Z_{n+1}$, Thus $-nx^{-1}Z_n + Z_n' = -Z_{n+1}$ and rearranging

$Z_n' = \frac{n}{x} Z_n - Z_{n+1}$ 1st order derivative

Consider the second one LHS $\frac{d}{dx}(x^n Z_n(x)) = nx^{n-1} Z_n + x^n Z_n'$ equating to RHS $nx^{n-1} Z_n + x^n Z_n' = x^n Z_{n-1}$, Thus $nx^{-1} Z_n + Z_n' = Z_{n-1}$ and rearranging

$Z_n' = -\frac{n}{x} Z_n + Z_{n-1}$ 1st order derivative, more of

On adding

$2Z_n' = Z_{n-1} - Z_{n+1}$ odd even indices

Returning to the first one $Z_n' = \frac{n}{x} Z_n - Z_{n+1}$ and multiplying by 2, in particular $2Z_n' = \frac{2n}{x} Z_n - 2Z_{n+1}$ and on comparing $Z_{n-1} - Z_{n+1} = \frac{2n}{x} Z_n - 2Z_{n+1}$, thus

$Z_{n-1} + Z_{n+1} = \frac{2n}{x} Z_n$ successive indices

Again returning to the first one $Z_n' = \frac{n}{x} Z_n - Z_{n+1}$ differentiating $Z_n'' = \frac{-n}{x^2} Z_n + \frac{n}{x} Z_n' - Z_{n+1}'$ and substituting the 1st order derivative back in $Z_n'' = \frac{-n}{x^2} Z_n + \frac{n}{x}(\frac{n}{x} Z_n - Z_{n+1}) - Z_{n+1}' = \frac{n(n-1)}{x^2} Z_n - \frac{n}{x} Z_{n+1} - Z_{n+1}'$

Using the second one $Z_n' = -\frac{n}{x} Z_n + Z_{n-1}$ for $n \leftarrow n+1$, in particular $Z_{n+1}' = -\frac{n+1}{x} Z_{n+1} + Z_n$, gets to $Z_n'' = \frac{n(n-1)}{x^2} Z_n - \frac{n}{x} Z_{n+1} - (-\frac{n+1}{x} Z_{n+1} + Z_n)$ that is $Z_n'' = \frac{n(n-1)}{x^2} Z_n + \frac{1}{x} Z_{n+1} - Z_n$

$Z_n'' = \left(\frac{n(n-1)}{x^2} - 1 \right) Z_n + \frac{1}{x} Z_{n+1}$ 2nd order derivative

§§§

General Field Computations

Field example

$curl\ \mathbf{H} = \mathbf{J} + j\omega\varepsilon\ \mathbf{E}$ $\quad$ $div\ \mathbf{H} = 0$

$curl\ \mathbf{E} = -j\omega\mu\ \mathbf{H}$ $\quad$ $div\ \mathbf{E} = \frac{\rho}{\varepsilon}$

$curl\ curl\ \mathbf{H} - k^2\mathbf{H} = curl\ \mathbf{J}$

$curl\ curl\ \mathbf{E} - k^2\mathbf{E} = -j\omega\mu\ \mathbf{J}$ where $k = \omega\sqrt{\varepsilon\mu}$

Let

$\mathbf{v} \equiv \mathbf{H}$, $\quad$ $\mathbf{u} \equiv \frac{e^{-jkr}}{r}\mathbf{b} = \psi\mathbf{b}$,

Notice that $\mathbf{b}$ is a const vector as Green's Theorem requires vector operands, and that $\psi \equiv \frac{e^{-jkr}}{r}$.
Also in anticipation of the integral over the sphere of radius r_0, i.e. over the point of interest (p.o.i), the normal vector $\mathbf{n}$ counts as positive when pointing inward. Thus $grad\,\psi = -(jk + \frac{1}{r})\frac{e^{-jkr}}{r}\mathbf{n}$

Green's Theorem:

$$\int_V (\mathbf{u}\ curl\ curl\ \mathbf{v} - \mathbf{v}\ curl\ curl\ \mathbf{u})dV = \oint_A (\mathbf{v} \times curl\ \mathbf{u} - \mathbf{u} \times curl\ \mathbf{v})d\mathbf{A}$$

Procedure: 1.) Introduce $\mathbf{J}$ (or ρ, ε) instead of the field vectors into the volume integral, 2.) Eliminate the arbitrary constant vector $\mathbf{b}$ from the equation, 3.) Then let the infinitesimal sphere of radius r_0 approach the point of interest.

$$\int_{V-G_0} (\frac{e^{-jkr}}{r}\mathbf{b}\ curl\ curl\ \mathbf{H} - \mathbf{H}\ curl\ curl\ \frac{e^{-jkr}}{r}\mathbf{b})dV = -\oint_{A_0+\sum_i A_i} (\mathbf{H} \times curl\ \frac{e^{-jkr}}{r}\mathbf{b} - \frac{e^{-jkr}}{r}\mathbf{b} \times curl\ \mathbf{H})d\mathbf{A}$$

The boundary surface A_N associated with the highest index encloses all other surfaces & sources and the entire space that encloses the p.o.i[4]

Before proceeding, recalling a few relationships from vector analysis.

---

[4] The derivation for the electric field $\mathbf{E}$ has been done whereelse, [e.g. Simonyi]. This text is following the same procedure to derive the expression for the magnetic field $\mathbf{H}$. Both expressions are shown in the Summary that follows.

About convention, the normal vector pointing away from the surface, i.e. pointing outward hitherto is taken to be positive. For this derivation however the point of interest is in an enclosed space, thus pointing inward counts as positive, opposite to the direction of the usual convention.

$$\mathbf{u}(\mathbf{v}\times\mathbf{w}) = \mathbf{w}(\mathbf{u}\times\mathbf{v}) = \mathbf{v}(\mathbf{w}\times\mathbf{u})$$
$$\mathbf{u}\times(\mathbf{v}\times\mathbf{w}) = (\mathbf{uw})\mathbf{v} - (\mathbf{uv})\mathbf{w}$$
$$(\mathbf{t}\times\mathbf{u})(\mathbf{v}\times\mathbf{w}) = \mathbf{t}[\mathbf{u}\times(\mathbf{v}\times\mathbf{w})]$$
$$(\mathbf{t}\times\mathbf{u})\times(\mathbf{v}\times\mathbf{w}) = [(\mathbf{t}\times\mathbf{u})\mathbf{w}]\mathbf{v} - [(\mathbf{t}\times\mathbf{u})\mathbf{v}]\mathbf{w}$$
$$div\ \varphi\mathbf{v} = \varphi\ div\ \mathbf{v} + \mathbf{v}\ grad\ \varphi$$
$$curl\ \varphi\mathbf{v} = \varphi\ curl\ \mathbf{v} + grad\ \varphi\times\mathbf{v}$$
$$div(\mathbf{u}\times\mathbf{v}) = \mathbf{v}curl\mathbf{u} - \mathbf{u}curl\mathbf{v}$$
$$\Delta(\mathbf{v}) \equiv grad\ div\ (\mathbf{v}) - curl\ curl\ (\mathbf{v})$$
$$curl\ grad\ \varphi = 0$$
$$div\ curl\ \mathbf{v} = 0$$

Looking @ the volume integral (on the LHS)

1st term: $curl\ curl\ \mathbf{H} = curl\ \mathbf{J} + k^2\mathbf{H}$

$\psi\mathbf{b}\ curl\ curl\ \mathbf{H} = \psi\mathbf{b}\ curl\ \mathbf{J} + \psi\mathbf{b}\ k^2\mathbf{H}$

2nd term: $curl\ curl\ (\psi\mathbf{b}) \equiv grad\ div\ (\psi\mathbf{b}) - \Delta(\psi\mathbf{b})$

where using $div\ \varphi\mathbf{v} = \varphi\ div\ \mathbf{v} + \mathbf{v}\ grad\ \varphi$ (with $\varphi \equiv \psi$, $\mathbf{v} \equiv \mathbf{b}$ const) thus $div\ \psi\mathbf{b} = \mathbf{b}\ grad\ \psi$

and $\Delta\psi + k^2\psi = 0$, $\Delta\psi = -k^2\psi$ thus $\Delta\psi\mathbf{b} = -k^2\psi\mathbf{b}$

$curl\ curl\ (\psi\mathbf{b}) \equiv grad\ (\mathbf{b}\ grad\ \psi) + k^2\psi\mathbf{b}$

$\mathbf{H}\ curl\ curl\ (\psi\mathbf{b}) \equiv \mathbf{H}\ grad\ (\mathbf{b}\ grad\ \psi) + \mathbf{H}\ k^2\psi\mathbf{b}$

$$\int_{V-G_0} (\cdots)dV = \int_{V-G_0} (\psi\mathbf{b}\ curl\ \mathbf{J} + \psi\mathbf{b}\ k^2\mathbf{H} - \mathbf{H}\ grad\ (\mathbf{b}\ grad\ \psi) - \mathbf{H}\ k^2\psi\mathbf{b})dV$$

$$\int_{V-G_0} (\cdots)dV = \int_{V-G_0} (\psi\mathbf{b}\ curl\ \mathbf{J} - \mathbf{H}\ grad\ (\mathbf{b}\ grad\ \psi))dV$$

from $div\ \varphi\mathbf{v} = \varphi\ div\ \mathbf{v} + \mathbf{v}\ grad\ \varphi$, $\mathbf{v}\ grad\ \varphi = div\ \varphi\mathbf{v} - \varphi\ div\ \mathbf{v}$

letting $\varphi = (\mathbf{b}\ grad\ \psi)$ and $\mathbf{v} = \mathbf{H}$

$\mathbf{H}\ grad\ (\mathbf{b}\ grad\ \psi) = div[(\mathbf{b}\ grad\ \psi)\mathbf{H}] - (\mathbf{b}\ grad\ \psi)(div\mathbf{H})$

but $div\ \mathbf{H} = 0$, then

$\mathbf{H}\ grad\ (\mathbf{b}\ grad\ \psi) = div[(\mathbf{b}\ grad\ \psi)\mathbf{H}]$

$$\int_{V-G_0} (\cdots)dV = \int_{V-G_0} (\psi\mathbf{b}\ curl\ \mathbf{J} - div[(\mathbf{b}\ grad\ \psi)\mathbf{H}])dV = \int_{V-G_0} (\psi\mathbf{b}\ curl\ \mathbf{J})dV - \int_{V-G_0} div[(\mathbf{b}\ grad\ \psi)\mathbf{H}]dV$$

Multiplied by minus one

$$-\int_{V-G_0} (\cdots)dV = -\int_{V-G_0} (\psi\mathbf{b}\ curl\ \mathbf{J})dV + \int_{V-G_0} div[(\mathbf{b}\ grad\ \psi)\mathbf{H}]dV$$

By Gauß theorem, also recalling normal vector pointing away from p.o.i. (into the space $V - G_0 \to V$)

$$\int_{V-G_0} div[(\mathbf{b}\ grad\ \psi)\mathbf{H}]dV = -\oint_{A_0+\sum_i A_i} (\mathbf{b}\ grad\ \psi)\mathbf{H}\ d\mathbf{A} = -\mathbf{b}\oint_{A_0+\sum_i A_i} (grad\ \psi)(\mathbf{H}\ d\mathbf{A})$$

$$-\int_{V-G_0} (\cdots)dV = -\int_{V-G_0} (\psi\mathbf{b}\ curl\ \mathbf{J})dV - \mathbf{b}\oint_{A_0+\sum_i A_i} (grad\ \psi)(\mathbf{H}\ d\mathbf{A})$$

$$-\int_{V-G_0}(\cdots)dV=\mathbf{b}\int_{V-G_0}(-\psi\ curl\ \mathbf{J})dV-\mathbf{b}\oint_{A_0+\sum_i A_i}(grad\ \psi)(\mathbf{H}\ d\mathbf{A})$$

The surface integral to be grouped together with all other surface integrals
Turning to the RHS multiplied by minus one

$$-\oint_{A_0+\sum_i A_i}(\cdots)d\mathbf{A}=\oint_{A_0+\sum_i A_i}(\mathbf{H}\times curl\frac{e^{-jkr}}{r}\mathbf{b}-\frac{e^{-jkr}}{r}\mathbf{b}\times curl\ \mathbf{H})d\mathbf{A}$$

1st term:

$curl\ \varphi\mathbf{v}=\varphi\ curl\ \mathbf{v}+grad\ \varphi\times\mathbf{v}$ (where letting $\varphi=\psi$ and $\mathbf{v}=\mathbf{b}$ const vector)

$curl\ \psi\mathbf{b}=\psi\ curl\ \mathbf{b}+grad\ \psi\times\mathbf{b}=grad\ \psi\times\mathbf{b}$

$\mathbf{u}(\mathbf{v}\times\mathbf{w})=\mathbf{w}(\mathbf{u}\times\mathbf{v})=\mathbf{v}(\mathbf{w}\times\mathbf{u})$

$(\mathbf{H}\times curl\ \psi\mathbf{b})d\mathbf{A}=d\mathbf{A}(\mathbf{H}\times curl\ \psi\mathbf{b})=curl\ \psi\mathbf{b}(d\mathbf{A}\times\mathbf{H})=(grad\ \psi\times\mathbf{b})(d\mathbf{A}\times\mathbf{H})$

$(grad\ \psi\times\mathbf{b})(d\mathbf{A}\times\mathbf{H})=-(\mathbf{b}\times grad\ \psi)(d\mathbf{A}\times\mathbf{H})$

$(\mathbf{t}\times\mathbf{u})(\mathbf{v}\times\mathbf{w})=\mathbf{t}[\mathbf{u}\times(\mathbf{v}\times\mathbf{w})]$

$-(\mathbf{b}\times grad\ \psi)(d\mathbf{A}\times\mathbf{H})=-\mathbf{b}[grad\ \psi\times(d\mathbf{A}\times\mathbf{H})]=\mathbf{b}[(d\mathbf{A}\times\mathbf{H})\times grad\ \psi]$

Thus $(\mathbf{H}\times curl\ \psi\mathbf{b})d\mathbf{A}=[(d\mathbf{A}\times\mathbf{H})\times grad\ \psi]\mathbf{b}$

2nd term:

$\psi[\mathbf{b}\times(\mathbf{J}+j\omega\varepsilon\ \mathbf{E})]d\mathbf{A}=$

$\psi[\mathbf{b}\times\mathbf{J}]d\mathbf{A}+\psi[\mathbf{b}\times(j\omega\varepsilon\ \mathbf{E})]d\mathbf{A}=\psi[\mathbf{b}\times\mathbf{J}]d\mathbf{A}-j\omega\varepsilon\psi[\mathbf{E}\times\mathbf{b}]d\mathbf{A}=\psi[\mathbf{b}\times\mathbf{J}]d\mathbf{A}-j\omega\varepsilon\psi[d\mathbf{A}\times\mathbf{E}]\mathbf{b}$

Assembling both sides

$$\int_{V-G_0}(-\psi\mathbf{b}\ curl\ \mathbf{J})dV-\mathbf{b}\oint_{A_0+\sum_i A_i}(grad\ \psi)(\mathbf{H}\ d\mathbf{A})=$$

$$\oint_{A_0+\sum_i A_i}[(d\mathbf{A}\times\mathbf{H})\times grad\ \psi]\mathbf{b}-\oint_{A_0+\sum_i A_i}\psi[\mathbf{b}\times\mathbf{J}]d\mathbf{A}+\oint_{A_0+\sum_i A_i}j\omega\varepsilon\psi[d\mathbf{A}\times\mathbf{E}]\mathbf{b}$$

On the RHS, converting the 2nd integral into a volume, again pointing away from p.o.i.

$$\int_{V-G_0}(-\psi\mathbf{b}\ curl\ \mathbf{J})dV-\mathbf{b}\oint_{A_0+\sum_i A_i}(grad\ \psi)(\mathbf{H}\ d\mathbf{A})=$$

$$\oint_{A_0+\sum_i A_i}[(d\mathbf{A}\times\mathbf{H})\times grad\ \psi]\mathbf{b}+\int_{V-G_0}div[\psi(\mathbf{b}\times\mathbf{J})]dV+\oint_{A_0+\sum_i A_i}j\omega\varepsilon\psi[d\mathbf{A}\times\mathbf{E}]\mathbf{b}$$

Grouping volume integrals onto the LHS, and surface integrals onto the RHS

$$\int_{V-G_0}(-\psi\mathbf{b}\ curl\ \mathbf{J})dV-\int_{V-G_0}div[\psi(\mathbf{b}\times\mathbf{J})]dV=$$

$$\oint_{A_0+\sum_i A_i}[(d\mathbf{A}\times\mathbf{H})\times grad\ \psi]\mathbf{b}+\oint_{A_0+\sum_i A_i}j\omega\varepsilon\psi[d\mathbf{A}\times\mathbf{E}]\mathbf{b}+\mathbf{b}\oint_{A_0+\sum_i A_i}(grad\ \psi)(\mathbf{H}\ d\mathbf{A})$$

Using $d\mathbf{A}=\mathbf{n}\ dA$

$$\int_{V-G_0}(-\psi\mathbf{b}\ curl\ \mathbf{J})dV-\int_{V-G_0}div[\psi(\mathbf{b}\times\mathbf{J})]dV=$$

$$\mathbf{b}\oint_{A_0+\sum_i A_i}[(\mathbf{n}\times\mathbf{H})\times grad\ \psi]\ dA+\mathbf{b}\oint_{A_0+\sum_i A_i}j\omega\varepsilon\psi[\mathbf{n}\times\mathbf{E}]\ dA+\mathbf{b}\oint_{A_0+\sum_i A_i}(\mathbf{n}\mathbf{H})(grad\ \psi)\ dA$$

On the LHS consider

$div\ \varphi\mathbf{v}=\varphi\ div\ \mathbf{v}+\mathbf{v}\ grad\ \varphi$

$div[\psi(\mathbf{b}\times\mathbf{J})] = \psi div(\mathbf{b}\times\mathbf{J}) + (\mathbf{b}\times\mathbf{J})grad\,\psi$

$div(\mathbf{u}\times\mathbf{v}) = \mathbf{v}curl\mathbf{u} - \mathbf{u}curl\mathbf{v}$

$div(\mathbf{b}\times\mathbf{J}) = \mathbf{J}curl\mathbf{b} - \mathbf{b}curl\mathbf{J} \quad = -\mathbf{b}curl\mathbf{J}$ as $\mathbf{b}$ constant

Then $div[\psi(\mathbf{b}\times\mathbf{J})] = \psi div(\mathbf{b}\times\mathbf{J}) + (\mathbf{b}\times\mathbf{J})grad\,\psi = -\psi\mathbf{b}curl\mathbf{J} + (\mathbf{b}\times\mathbf{J})grad\,\psi$

$$\int_{V-G_0} -(\mathbf{b}\times\mathbf{J})grad\,\psi\; dV =$$

$$\mathbf{b}\oint_{A_0+\sum_i A_i} [(\mathbf{n}\times\mathbf{H})\times grad\;\psi]\, dA + \mathbf{b}\oint_{A_0+\sum_i A_i} j\omega\varepsilon\psi[\mathbf{n}\times\mathbf{E}]\, dA + \mathbf{b}\oint_{A_0+\sum_i A_i} (\mathbf{nH})(grad\;\psi)\, dA$$

Because $-(\mathbf{b}\times\mathbf{J})grad\,\psi = (\mathbf{J}\times\mathbf{b})grad\,\psi = grad\,\psi(\mathbf{J}\times\mathbf{b}) = \mathbf{b}(grad\,\psi\times\mathbf{J}) = -\mathbf{b}(\mathbf{J}\times grad\,\psi)$

$$-\mathbf{b}\int_{V-G_0} (\mathbf{J}\times grad\,\psi)\; dV =$$

$$\mathbf{b}\oint_{A_0+\sum_i A_i} [(\mathbf{n}\times\mathbf{H})\times grad\;\psi]\, dA + \mathbf{b}\oint_{A_0+\sum_i A_i} j\omega\varepsilon\psi[\mathbf{n}\times\mathbf{E}]\, dA + \mathbf{b}\oint_{A_0+\sum_i A_i} (\mathbf{nH})(grad\;\psi)\, dA$$

Dropping the arbitrary const vector $\mathbf{b}$

$$-\int_{V-G_0} (\mathbf{J}\times grad\,\psi)\; dV = \oint_{A_0+\sum_i A_i} \left(j\omega\varepsilon\psi(\mathbf{n}\times\mathbf{E}) + [(\mathbf{n}\times\mathbf{H})\times grad\;\psi] + (\mathbf{nH})grad\;\psi\right) dA$$

Isolating the p.o.i to one side of the expression

$$\oint_{A_0} \left(j\omega\varepsilon\psi(\mathbf{n}\times\mathbf{E}) + [(\mathbf{n}\times\mathbf{H})\times grad\;\psi] + (\mathbf{nH})grad\;\psi\right) dA =$$

$$-\int_{V-G_0} (\mathbf{J}\times grad\,\psi)\; dV - \oint_{\sum_i A_i} \left(j\omega\varepsilon\psi(\mathbf{n}\times\mathbf{E}) + [(\mathbf{n}\times\mathbf{H})\times grad\;\psi] + (\mathbf{nH})grad\;\psi\right) dA$$

Before letting $r_0 \to 0$, recalling that $grad\,\psi\,|_{r=r_0} = -(jk + \frac{1}{r_0})\frac{e^{-jkr_0}}{r_0}\mathbf{n}$

Proceeding to let $r_0 \to 0$

$$\oint_{A_0} \left(j\omega\varepsilon\psi(\mathbf{n}\times\mathbf{E}) + [(\mathbf{n}\times\mathbf{H})\times grad\;\psi] + (\mathbf{nH})grad\;\psi\right) dA =$$

$$\oint_{A_0} \left(j\omega\varepsilon\psi(\mathbf{n}\times\mathbf{E}) + [(\mathbf{n}\times\mathbf{H})\times\left(-(jk+\frac{1}{r_0})\frac{e^{-jkr_0}}{r_0}\mathbf{n}\right)] + (\mathbf{nH})\left(-(jk+\frac{1}{r_0})\frac{e^{-jkr_0}}{r_0}\mathbf{n}\right)\right) dA =$$

writing $dA = r_0^2 d\Omega$

$$= \oint_{\Omega} \left(j\omega\varepsilon\frac{e^{-jkr_0}}{r_0}(\mathbf{n}\times\mathbf{E}) + [(\mathbf{n}\times\mathbf{H})\times\left(-(jk+\frac{1}{r_0})\frac{e^{-jkr_0}}{r_0}\mathbf{n}\right)] + (\mathbf{nH})\left(-(jk+\frac{1}{r_0})\frac{e^{-jkr_0}}{r_0}\mathbf{n}\right)\right) r_0^2 d\Omega$$

$$= r_0 e^{-jkr_0}\oint_{\Omega} \left(j\omega\varepsilon(\mathbf{n}\times\mathbf{E}) + [(\mathbf{n}\times\mathbf{H})\times\left(-(jk+\frac{1}{r_0})\mathbf{n}\right)] + (\mathbf{nH})\left(-(jk+\frac{1}{r_0})\mathbf{n}\right)\right) d\Omega$$

$$= jr_0 e^{-jkr_0}\oint_{\Omega} \omega\varepsilon(\mathbf{n}\times\mathbf{E})\, d\Omega - jr_0 e^{-jkr_0}\oint_{\Omega} [(\mathbf{n}\times\mathbf{H})\times k\mathbf{n} + (\mathbf{nH})k\mathbf{n}]\, d\Omega - e^{-jkr_0}\oint_{\Omega} [(\mathbf{n}\times\mathbf{H})\times\mathbf{n} + (\mathbf{nH})\mathbf{n}]\, d\Omega$$

from $(\mathbf{n}\times\mathbf{H})\times\mathbf{n} = -\mathbf{n}\times(\mathbf{n}\times\mathbf{H}) = -(\mathbf{nH})\mathbf{n} + (\mathbf{nn})\mathbf{H}$, the identity $\mathbf{H} = (\mathbf{n}\times\mathbf{H})\times\mathbf{n} + (\mathbf{nH})\mathbf{n}$

$$= jr_0 e^{-jkr_0}\oint_{\Omega} \omega\varepsilon(\mathbf{n}\times\mathbf{E})\, d\Omega - jr_0 e^{-jkr_0}\oint_{\Omega} k\mathbf{H}\, d\Omega - e^{-jkr_0}\oint_{\Omega} \mathbf{H}\, d\Omega$$

the first 2 terms approach zero, leaving the remaining term approaching the p.o.i, where $V - G_0 \to V$

$$\lim_{r_0 \to 0} \oint_{A_0} (\cdots) dA = 0 - 4\pi \mathrm{e}^{-jkr_0} \mathbf{H}_{average}$$

For the p.o.i, thus

$$-4\pi\mathbf{H} = -\int_V (\mathbf{J} \times grad\psi)\, dV - \oint_{\sum_i A_i} \left(j\omega\varepsilon\psi(\mathbf{n}\times\mathbf{E}) + [(\mathbf{n}\times\mathbf{H}) \times grad\ \psi] + (\mathbf{nH}) grad\ \psi \right) dA$$

$$\mathbf{H} = \frac{1}{4\pi}\int_V (\mathbf{J} \times grad\psi)\, dV + \frac{1}{4\pi}\oint_{\sum_i A_i} \left(j\omega\varepsilon(\mathbf{n}\times\mathbf{E})\psi + [(\mathbf{n}\times\mathbf{H}) \times grad\ \psi] + (\mathbf{nH}) grad\ \psi \right) dA$$

That is

$$\mathbf{H} = \frac{1}{4\pi}\int_V (\mathbf{J} \times grad\frac{\mathrm{e}^{-jkr}}{r})\, dV + \frac{1}{4\pi}\oint_{\sum_i A_i} \left(j\omega\varepsilon(\mathbf{n}\times\mathbf{E})\frac{\mathrm{e}^{-jkr}}{r} + [(\mathbf{n}\times\mathbf{H}) \times grad\ \frac{\mathrm{e}^{-jkr}}{r}] + (\mathbf{nH}) grad\ \frac{\mathrm{e}^{-jkr}}{r} \right) dA$$

Conditions for propagation

Consider the surface integral associated with A_N and recall $grad\psi = -(jk + \frac{1}{r})\frac{\mathrm{e}^{-jkr}}{r}\mathbf{n}$

$$\frac{1}{4\pi}\oint_{A_N} \left(j\omega\varepsilon(\mathbf{n}\times\mathbf{E})\frac{\mathrm{e}^{-jkr}}{r} + [(\mathbf{n}\times\mathbf{H}) \times grad\ \frac{\mathrm{e}^{-jkr}}{r}] + (\mathbf{nH}) grad\ \frac{\mathrm{e}^{-jkr}}{r} \right) dA$$

$$= \frac{1}{4\pi}\oint_{A_N} \left(j\omega\varepsilon(\mathbf{n}\times\mathbf{E})\frac{\mathrm{e}^{-jkr}}{r} + [(\mathbf{n}\times\mathbf{H}) \times \left(-(jk + \frac{1}{r})\frac{\mathrm{e}^{-jkr}}{r}\mathbf{n} \right)] + (\mathbf{nH})\left(-(jk + \frac{1}{r})\frac{\mathrm{e}^{-jkr}}{r}\mathbf{n} \right) \right) dA$$

$$= \frac{1}{4\pi}\oint_{A_N} \left(j\omega\varepsilon(\mathbf{n}\times\mathbf{E}) + [(\mathbf{n}\times\mathbf{H}) \times \left(-(jk + \frac{1}{r})\mathbf{n} \right)] + (\mathbf{nH})\left(-(jk + \frac{1}{r})\mathbf{n} \right) \right) \frac{\mathrm{e}^{-jkr}}{r} dA$$

$$\ldots -(jk + \frac{1}{r})\{(\mathbf{n}\times\mathbf{H}) \times (\mathbf{n}) \quad + \quad (\mathbf{nH})(\mathbf{n})\}\ldots$$

$$\ldots -(jk + \frac{1}{r})\{-(\mathbf{n}) \times (\mathbf{n}\times\mathbf{H}) + \quad (\mathbf{nH})(\mathbf{n})\}\ldots$$

$$= \frac{1}{4\pi}\oint_{A_N} \left(j\omega\varepsilon(\mathbf{n}\times\mathbf{E}) \quad - \quad (jk + \frac{1}{r})\{-(\mathbf{n}) \times (\mathbf{n}\times\mathbf{H}) + \quad (\mathbf{nH})(\mathbf{n})\} \right) \frac{\mathrm{e}^{-jkr}}{r} dA$$

expanding A_N to infinity, also the normal vector to point outward into infinity, $\mathbf{r}^N = -\mathbf{n}$, thus writing

$$= \frac{1}{4\pi}\oint_{A_N} \left(-j\omega\varepsilon(\mathbf{n}\times\mathbf{E}) \quad -(jk + \frac{1}{r})\{(\mathbf{n}) \times (\mathbf{n}\times\mathbf{H}) \quad - \quad (\mathbf{nH})(\mathbf{n})\} \right) \frac{\mathrm{e}^{-jkr}}{r} (-dA)$$

$$= \frac{1}{4\pi}\oint_{A_N} \left(-j\omega\varepsilon(\mathbf{r}^N\times\mathbf{E}) \quad -(jk + \frac{1}{r})\{(\mathbf{r}^N) \times (\mathbf{r}^N\times\mathbf{H}) \quad - \quad (\mathbf{r}^N\mathbf{H})(\mathbf{r}^N)\} \right) \frac{\mathrm{e}^{-jkr}}{r} dA$$

$$= \frac{1}{4\pi}\oint_{A_N} \left(-j\omega\varepsilon(\mathbf{r}^N\times\mathbf{E}) \quad +(jk + \frac{1}{r})\{(\mathbf{r}^N\mathbf{H})(\mathbf{r}^N) \quad - \quad (\mathbf{r}^N) \times (\mathbf{r}^N\times\mathbf{H})\} \right) \frac{\mathrm{e}^{-jkr}}{r} dA$$

Recalling $\mathbf{u} \times (\mathbf{v} \times \mathbf{w}) = (\mathbf{uw})\mathbf{v} - (\mathbf{uv})\mathbf{w}$, thus from $\mathbf{H} = (\mathbf{r}^N\mathbf{H})(\mathbf{r}^N) \quad - \quad (\mathbf{r}^N) \times (\mathbf{r}^N \times \mathbf{H})$

$$= \frac{1}{4\pi}\oint_{A_N} \left(-j\omega\varepsilon(\mathbf{r}^N\times\mathbf{E}) \quad +(jk + \frac{1}{r})\mathbf{H} \right) \frac{\mathrm{e}^{-jkr}}{r} dA$$

$$= \frac{1}{4\pi}\oint_{A_N} \left(-j\omega\varepsilon(\mathbf{r}^N\times\mathbf{E}) \quad + \quad jk\mathbf{H} + \frac{\mathbf{H}}{r} \right) \frac{\mathrm{e}^{-jkr}}{r} dA$$

$$= \frac{1}{4\pi} \oint_{A_N} \left(j\omega\varepsilon \left\{ -(\mathbf{r}^N \times \mathbf{E}) + \frac{k}{\omega\varepsilon}\mathbf{H} \right\} \quad + \quad \frac{\mathbf{H}}{r} \right) \frac{\mathrm{e}^{-jkr}}{r} \, dA$$

recalling $k = \omega\sqrt{\mu\varepsilon}$, i.e. $\frac{k}{\omega\varepsilon} = \frac{\omega\sqrt{\mu\varepsilon}}{\omega\varepsilon} = \sqrt{\frac{\mu}{\varepsilon}}$

$$= \frac{1}{4\pi} \oint_{A_N} \left(j\omega\varepsilon \left\{ -(\mathbf{r}^N \times \mathbf{E}) + \sqrt{\frac{\mu}{\varepsilon}}\mathbf{H} \right\} \quad + \quad \frac{\mathbf{H}}{r} \right) \frac{\mathrm{e}^{-jkr}}{r} \, dA$$

This integral vanishes for $\lim_{r\to\infty} r\mathbf{H} < const$ (i.e. $\mathbf{H}$ going as fast as $\frac{const}{r}$ for $r \to \infty$) and $\lim_{r\to\infty} \left\{ -(\mathbf{r}^N \times \mathbf{E}) + \sqrt{\frac{\mu}{\varepsilon}}\mathbf{H} \right\} = 0$, the condition for propagation thus reading as $-(\mathbf{r}^N \times \mathbf{E}) + \sqrt{\frac{\mu}{\varepsilon}}\mathbf{H} \to 0$, accordingly $\mathbf{H} \to \sqrt{\frac{\varepsilon}{\mu}}(\mathbf{r}^N \times \mathbf{E})$.

Likewise, starting from the expression for $\mathbf{E}$, the condition would have been arrived at as $\lim_{r\to\infty} r\mathbf{E} < const$ and $\mathbf{E} \to -\sqrt{\frac{\mu}{\varepsilon}}(\mathbf{r}^N \times \mathbf{H})$.

At large distances away from all sources, both $\mathbf{H}$ and $\mathbf{E}$ are thus to behave like plane waves.

Summary

$$\mathbf{H} = \frac{1}{4\pi}\int_V (\mathbf{J}\times grad\frac{e^{-jkr}}{r})\,dV + \frac{1}{4\pi}\oint_{\sum_i A_i}\left(j\omega\varepsilon(\mathbf{n}\times\mathbf{E})\frac{e^{-jkr}}{r} + [(\mathbf{n}\times\mathbf{H})\times grad\,\frac{e^{-jkr}}{r}] + (\mathbf{nH})grad\,\frac{e^{-jkr}}{r}\right)dA$$

$$\mathbf{E} = \frac{1}{4\pi}\int_V \left(-j\omega\mu\mathbf{J}\frac{e^{-jkr}}{r} + \frac{\rho}{\varepsilon}grad\frac{e^{-jkr}}{r}\right)dV + \frac{1}{4\pi}\oint_{\sum_i A_i}\left(-j\omega\mu(\mathbf{n}\times\mathbf{H})\frac{e^{-jkr}}{r} + [(\mathbf{n}\times\mathbf{E})\times grad\,\frac{e^{-jkr}}{r}] + (\mathbf{nE})grad\,\frac{e^{-jkr}}{r}\right)dA$$

Scattering, Diffraction, Huygens' source, General problems of Classical Physics

Boundary surface at ∞, Local sources

$$\mathbf{H} = \frac{1}{4\pi}\int_V (\mathbf{J}\times grad\frac{e^{-jkr}}{r})\,dV$$

$$\mathbf{E} = \frac{1}{4\pi}\int_V \left(-j\omega\mu\mathbf{J}\frac{e^{-jkr}}{r} + \frac{\rho}{\varepsilon}grad\frac{e^{-jkr}}{r}\right)dV$$

$\mathbf{H} = curl\,\mathbf{A}$, equivalently, alternatively

$\mathbf{E} = -\mu\dfrac{\partial\mathbf{A}}{\partial t} - grad\;\varphi$, equivalently, alternatively

Electrostatics
Static dipoles

Antennas
Elementary dipol antennas,
Linear antennas

Boundary surface at ∞, Local potential surfaces

$$\mathbf{H} = \frac{1}{4\pi}\oint_{\sum_i A_i}\left(j\omega\varepsilon(\mathbf{n}\times\mathbf{E})\frac{e^{-jkr}}{r} + [(\mathbf{n}\times\mathbf{H})\times grad\,\frac{e^{-jkr}}{r}] + (\mathbf{nH})grad\,\frac{e^{-jkr}}{r}\right)dA$$

$$\mathbf{E} = \frac{1}{4\pi}\oint_{\sum_i A_i}\left(-j\omega\mu(\mathbf{n}\times\mathbf{H})\frac{e^{-jkr}}{r} + [(\mathbf{n}\times\mathbf{E})\times grad\,\frac{e^{-jkr}}{r}] + (\mathbf{nE})grad\,\frac{e^{-jkr}}{r}\right)dA$$

67 of 84

$$grad\,\phi - curl\,curl\,\mathbf{\Pi} + k^2\mathbf{\Pi} = 0$$

$$\mathbf{\Pi} = (\Pi_1, 0, 0)\,;\; g_1 = 1\,;\; \frac{\partial g_2 / g_3}{\partial x_1} \equiv 0\,;\; k^2 = -j\omega\mu(\sigma + j\omega\varepsilon)$$

	General orthogonal coordinates	Cylindrical coordinates	Spherical coordinates
	$x_1, x_2, x_3;\; g_1, g_2, g_3\; (ds_i = g_i dx_i);\; \mathbf{\Pi} = (\Pi_1, 0, 0)$	$z, r, \varphi;\; 1, 1, r;\; \mathbf{\Pi} = (\Pi_z, 0, 0);\; \phi = div\,\mathbf{\Pi}$	$r, \theta, \varphi;\; 1, r, r\sin\theta;\; \mathbf{\Pi} = (\Pi_r, 0, 0);\; \phi = \frac{1}{g_1}\frac{\partial \Pi_1}{\partial x_1}$
	$\frac{\partial^2 \Pi_1}{\partial x_1^2} + \frac{1}{g_2 g_3}\frac{\partial}{\partial x_2}\left(\frac{g_3}{g_2}\frac{\partial \Pi_1}{\partial x_2}\right) + \frac{1}{g_2 g_3}\frac{\partial}{\partial x_3}\left(\frac{g_2}{g_3}\frac{\partial \Pi_1}{\partial x_3}\right) + k^2 \Pi_1 = 0$	$\frac{\partial^2 \Pi_z}{\partial z^2} + \frac{1}{r}\frac{\partial}{\partial r}\left(r\frac{\partial \Pi_z}{\partial r}\right) + \frac{1}{r^2}\frac{\partial^2 \Pi_z}{\partial \varphi^2} + k^2 \Pi_z = 0$	$\frac{\partial^2 \Pi_r}{\partial r^2} + \frac{1}{r^2 \sin\theta}\frac{\partial}{\partial \theta}\left(\sin\theta \frac{\partial \Pi_r}{\partial \theta}\right) + \frac{1}{r^2 \sin^2\theta}\frac{\partial^2 \Pi_r}{\partial \varphi^2} + k^2 \Pi_r = 0$
	$\Pi_1(x_1, x_2, x_3; t) = X_1(x_1) X_2(x_2) X_3(x_3) \mathrm{e}^{j\omega t}$	$\Pi_z(z, r, \varphi; t) = Z_m\left(\sqrt{k^2 - \beta^2}\, r\right) \mathrm{e}^{\pm jm\varphi} \mathrm{e}^{\pm j\beta z} \mathrm{e}^{j\omega t}$	$\Pi_r(r, \theta, \varphi; t) = \sqrt{kr}\, Z_{n+\frac{1}{2}}(kr) P_n^m(\cos\theta) \mathrm{e}^{\pm jm\varphi} \mathrm{e}^{j\omega t}$
TM $\mathbf{H} = (\sigma + j\varepsilon\omega)\, curl\,\mathbf{\Pi}\,;\; \mathbf{E} = k^2\,\mathbf{\Pi} + grad\,\phi$			
	$E_1 = k^2 \Pi_1 + \frac{\partial^2 \Pi_1}{\partial x_1^2}$ $H_1 = 0$	$E_z = (k^2 - \beta^2)\Pi_z$ $H_z = 0$	$E_r = k^2 \Pi_r + \frac{\partial^2 \Pi_r}{\partial r^2}$ $H_r = 0$
	$E_2 = \frac{1}{g_2}\frac{\partial^2 \Pi_1}{\partial x_1 \partial x_2}$ $H_2 = \frac{\sigma + j\omega\varepsilon}{g_3}\frac{\partial \Pi_1}{\partial x_3}$	$E_r = -j\beta\frac{\partial \Pi_z}{\partial r}$ $H_r = \frac{jk^2}{\omega\mu r}\frac{\partial \Pi_z}{\partial \varphi}$	$E_\theta = \frac{1}{r}\frac{\partial^2 \Pi_r}{\partial r \partial \theta}$ $H_\theta = \frac{\sigma + j\omega\varepsilon}{r\sin\theta}\frac{\partial \Pi_r}{\partial \varphi}$
	$E_3 = \frac{1}{g_3}\frac{\partial^2 \Pi_1}{\partial x_1 \partial x_3}$ $H_3 = -\frac{\sigma + j\omega\varepsilon}{g_2}\frac{\partial \Pi_1}{\partial x_2}$	$E_\varphi = -j\frac{\beta}{r}\frac{\partial \Pi_z}{\partial \varphi}$ $H_\varphi = -\frac{jk^2}{\omega\mu}\frac{\partial \Pi_z}{\partial r}$	$E_\varphi = \frac{1}{r\sin\theta}\frac{\partial^2 \Pi_r}{\partial r \partial \varphi}$ $H_\varphi = -\frac{\sigma + j\omega\varepsilon}{r}\frac{\partial \Pi_r}{\partial \theta}$
TE $\mathbf{E} = -j\mu\omega\, curl\,\mathbf{\Pi}\,;\; \mathbf{H} = k^2\,\mathbf{\Pi} + grad\,\phi$			
	$E_1 = 0$ $H_1 = k^2 \Pi_1 + \frac{\partial^2 \Pi_1}{\partial x_1^2}$	$E_z = 0$ $H_z = (k^2 - \beta^2)\Pi_z$	$E_r = 0$ $H_r = k^2 \Pi_r + \frac{\partial^2 \Pi_r}{\partial r^2}$
	$E_2 = \frac{-j\omega\mu}{g_3}\frac{\partial \Pi_1}{\partial x_3}$ $H_2 = \frac{1}{g_2}\frac{\partial^2 \Pi_1}{\partial x_1 \partial x_2}$	$E_r = -\frac{j\omega\mu}{r}\frac{\partial \Pi_z}{\partial \varphi}$ $H_r = -j\beta\frac{\partial \Pi_z}{\partial r}$	$E_\theta = -\frac{\mu j\omega}{r\sin\theta}\frac{\partial \Pi_r}{\partial \varphi}$ $H_\theta = \frac{1}{r}\frac{\partial^2 \Pi_r}{\partial r \partial \theta}$
	$E_3 = \frac{j\omega\mu}{g_2}\frac{\partial \Pi_1}{\partial x_2}$ $H_3 = \frac{1}{g_3}\frac{\partial^2 \Pi_1}{\partial x_1 \partial x_3}$	$E_\varphi = j\omega\mu\frac{\partial \Pi_z}{\partial r}$ $H_\varphi = -j\frac{\beta}{r}\frac{\partial \Pi_z}{\partial \varphi}$	$E_\varphi = \frac{\mu j\omega}{r}\frac{\partial \Pi_r}{\partial \theta}$ $H_\varphi = \frac{1}{r\sin\theta}\frac{\partial^2 \Pi_r}{\partial r \partial \varphi}$

[Simonyi]

§§§

68 of 84

Miscellaneous

The Green function

The wave equations shown hitherto may be seen as particular cases of the equation

$\mathcal{D}\psi = f$

where $\mathcal{D}$ is a linear differential operator. In particular

$$\mathcal{D} \equiv \sum_{i,k} A_{ik} \frac{\partial^2}{\partial x_i \partial x_k} + \sum_{i} B_{ik} \frac{\partial}{\partial x_i} + C,$$

i.e.

$$\mathcal{D}\psi \equiv \sum_{i,k} A_{ik} \frac{\partial^2 \psi}{\partial x_i \partial x_k} + \sum_{i} B_{ik} \frac{\partial \psi}{\partial x_i} + C\psi.$$

The Green function may be seen as solution to

$\mathcal{D}G = I$

$\{I\} = \delta(\mathbf{x} - \mathbf{x'})$ being the kernel associated with the identity operator I, $\mathbf{x}$ the generalized variable along with the integration variable $\mathbf{x'}$, and G an integral operator representing the Green function.

One word about notations below[5]

One is tempted to write from $\mathcal{D}G = I$

$G = \mathcal{D}^{-1}$

Assuming G is known, operating G on its operand f i.e.

$\psi = \mathcal{D}^{-1} f = Gf$

$$\psi(\mathbf{x}) = \int_V G(\mathbf{x}, \mathbf{x'})\, f(\mathbf{x'})\, dV'$$

would yield the solution ψ.

Integral equation method

It is interesting to consider the homogeneous case of $\mathcal{D}_h$

$\mathcal{D}_h\psi \equiv \mathcal{D}\psi - \lambda\psi = 0$

Assuming the Green function associated with $\mathcal{D}$ of $\mathcal{D}\psi - \lambda\psi = 0$ is known, i.e. $\mathcal{D}G = I$, operate both sides on $\lambda\psi$ (multiply both sides with $\lambda\psi$ and integrate over V) to have

$$\mathcal{D} \int_V G(\mathbf{x}, \mathbf{x'})\, \lambda\psi(\mathbf{x'})\, dV' = \lambda\psi(\mathbf{x})$$

But this is another form of $\mathcal{D}\psi = f$, where

$$\psi(\mathbf{x}) = \int_V G(\mathbf{x}, \mathbf{x'})\, \lambda\psi(\mathbf{x'})\, dV' \qquad \text{and} \qquad f = \lambda\psi(\mathbf{x})$$

Rewriting the first expression as

$$\psi(\mathbf{x}) - \lambda \int_V G(\mathbf{x}, \mathbf{x'})\psi(\mathbf{x'})\, dV' = 0$$

[5] An integral operator written next to a differential operator $\mathcal{D}G = I$, mathematicians would require that ranges and domains for the operators be addressed first before writing an expression like that.

shows an integral equation for $\psi(\mathbf{x})$ that can exist only for certain λ-values, the problem known as the eigenvalue problem. The boundary conditions in general can only be satisfied for those λ-values.

Assumming self-adjoint differential operator, every function ψ that satisfies the boundary conditions may be represented as a linear combination of the eigenfunctions in the form

$$\psi = \sum_i a_i \psi_i ,$$

which is the most important result. The coefficients a_i are determined from the property that the eigenfunctions ψ_i are orthogonal to each other.

Addendum

Generic expressions

Eq.	Finite space (bounded by A)	inFinite space	
$\Delta\psi - \frac{1}{v^2}\frac{\partial^2}{\partial t^2}\psi = -g(\mathbf{r})$	$\psi(\mathbf{r},t) = \frac{1}{4\pi}\int_V \frac{[g]}{r}dV +$ $\frac{1}{4\pi}\int_A \left(\frac{1}{r}\left[\frac{\partial\psi}{\partial n}\right] - [\psi]\frac{\partial}{\partial n}\frac{1}{r} + \frac{1}{vr}\frac{\partial r}{\partial n}\left[\frac{\partial\psi}{\partial t}\right]\right)dA$ Note 1	$\psi(\mathbf{r},t) = \frac{1}{4\pi}\int_V \frac{g(\mathbf{r},t\mp\frac{r}{v})}{r}dV$	EMF
	$\psi(\mathbf{r},t) = \frac{1}{4\pi}\int_A \left(\frac{\partial\psi}{\partial n}\frac{e^{-j\mathbf{kr}}}{r} - \psi\frac{\partial}{\partial n}\frac{e^{-j\mathbf{kr}}}{r}\right)dA$ Note 2, $\psi = \psi_0\, e^{j(\omega t - \mathbf{kr})}$, $g \equiv 0$ assumed		Optics
$\Delta\psi = -g(\mathbf{r})$	$\psi(\mathbf{r},t) = -\frac{1}{4\pi}\int_V \frac{\Delta\psi}{r}dV +$ $\frac{1}{4\pi}\int_A \frac{\partial\psi}{\partial n}\frac{1}{r}dA - \frac{1}{4\pi}\int_A \psi\frac{\partial}{\partial n}\frac{1}{r}dA$	$\psi(\mathbf{r},t) = \frac{1}{4\pi}\int_V \frac{-\Delta\psi}{r}dV$	Electro statics
$\Delta\psi - \frac{1}{v^2}\frac{\partial^2}{\partial t^2}\psi = 0$	$\psi_0 f(t \mp \frac{\mathbf{nr}}{v})$		

1. The square brackets denotes evaluation to be taken at $t \mp \frac{r}{v}$ on having completed the operations shown
2. Multiplying $t \mp \frac{r}{v}$ by ω so as to write $\omega t \mp \frac{\omega}{v} r$, or $\omega t \mp kr$ where $k = \frac{\omega}{v}$, the harmonic time dependency assumption leads to the relationship shown (Kirchhoff Eq.)

The generic expressions represent a bottleneck or are unusable as ψ and $\frac{\partial}{\partial n}\psi$ on A could not be specified independently from each other (ψ is defined, once either ψ or $\frac{\partial}{\partial n}\psi$ on A has been specified). Obtaining $\mathbf{E}$ and $\mathbf{H}$, related not only by the wave Eqs but also by the Maxwell Eqs, would not be a simple task. Selecting solutions that correspond to each other would be another challenge.

4-dim spacetime formalism

	Eq.		
Lagrange equation of the 2nd kind	$L \equiv F(t,q_i,\dot{q}_i)$, base function or base integrand $\frac{\partial L}{\partial q_i}-\frac{d}{dt}\frac{\partial L}{\partial \dot{q}_i}=0$, $i=1,2,3,\cdots,f$	Generalized coordinates q_i for a system with f degrees of freedom, $L=W_{kinetic}-W_{potential}$	Formal language of mechanics
Spatial (one-dim) time derivatives extension for the Euler Eq.	$\mathcal{L}\equiv F(t,q,\dot{q},q')$ $\frac{\partial \mathcal{L}}{\partial q}-\frac{\partial}{\partial x}\left[\frac{\partial \mathcal{L}}{\partial(\frac{\partial q}{\partial x})}\right]-\frac{\partial}{\partial t}\left[\frac{\partial \mathcal{L}}{\partial(\frac{\partial q}{\partial t})}\right]=0$	The q_i replaced by their distributed equivalent q	
4-dim spacetime formalism	$\mathcal{L}\equiv F(t,\eta,\frac{\partial \eta}{\partial t},grad\,\eta)$	$\eta(x,y,z,t)$ is sought such that $\delta\int_V \mathcal{L}\,dV=0$	
	$\frac{\partial \mathcal{L}}{\partial \eta}-\sum_{k=1}^{3}\frac{\partial}{\partial x_k}\left[\frac{\partial \mathcal{L}}{\partial(\frac{\partial \eta}{\partial x_k})}\right]-\frac{\partial}{\partial t}\left[\frac{\partial \mathcal{L}}{\partial(\frac{\partial \eta}{\partial t})}\right]=0$	Identical to the original $\frac{\partial L}{\partial q_i}-\frac{d}{dt}\frac{\partial L}{\partial \dot{q}_i}=0$ on defining	
	$\frac{\delta \mathcal{L}}{\delta \eta}-\frac{\partial}{\partial t}\left[\frac{\partial \mathcal{L}}{\partial(\frac{\partial \eta}{\partial t})}\right]=0$	$\frac{\delta \mathcal{L}}{\delta \eta}\equiv\frac{\partial \mathcal{L}}{\partial \eta}-\sum_{k=1}^{3}\frac{\partial}{\partial x_k}\left[\frac{\partial \mathcal{L}}{\partial(\frac{\partial \eta}{\partial x_k})}\right]$	
Hamilton function	$H=\int_V \mathcal{H}\,dV$ $\mathcal{H}=\mathcal{P}\dot{\eta}-\mathcal{L}$, $\mathcal{P}=\frac{\partial \mathcal{L}}{\partial \dot{\eta}}$	$\mathcal{H}$: Hamilton density funct, $\mathcal{P}$: impulse density $\dot{\mathcal{P}}=-\frac{\delta \mathcal{H}}{\delta \eta}$, $\dot{\eta}=\frac{\delta \mathcal{H}}{\delta \mathcal{P}}$	
Lagange density funct of EMF	$\mathcal{L}=\frac{1}{2}\varepsilon(\frac{\partial \mathbf{A}}{\partial t}+grad\varphi)^2-\frac{1}{2\mu}(curl\,\mathbf{A})^2$	$\mathbf{E}=-\frac{\partial \mathbf{A}}{\partial t}-grad\varphi$, $\mathbf{B}=curl\,\mathbf{A}$	Maxwell Eqs. in the language of mechanics
4-dim spacetime formalism	$\frac{\partial \mathcal{L}}{\partial A_\alpha}-\sum_{k=1}^{3}\frac{\partial}{\partial x_k}\left[\frac{\partial \mathcal{L}}{\partial(\frac{\partial A_\alpha}{\partial x_k})}\right]-\frac{\partial}{\partial t}\left[\frac{\partial \mathcal{L}}{\partial(\frac{\partial A_\alpha}{\partial t})}\right]=0$ $\alpha=1,2,3,4$	$A_1,A_2,A_3,A_4\equiv j\varphi/c$ being treated as state variables. On carrying out the differentiations, the expression shows itself equivalent to the Maxwell Equations I II III IV for $\mathbf{J}=0$ and $\rho=0$	
Hamilton density funct of EMF	$\mathcal{L}=\frac{1}{2}\varepsilon(\dot{\eta}+grad\varphi)^2-\frac{1}{2\mu}(curl\,\eta)^2$ $\mathcal{H}=\mathcal{P}\dot{\eta}-\mathcal{L}=\mathcal{P}\frac{\mathcal{P}}{\varepsilon}-\mathcal{L}$ $\mathcal{H}=\frac{\mathcal{P}^2}{\varepsilon}-\left[\frac{1}{2}\varepsilon(\dot{\eta})^2-\frac{1}{2\mu}(curl\,\eta)^2\right]$ $\mathcal{H}=\frac{\mathcal{P}^2}{\varepsilon}-\left[\frac{1}{2}\varepsilon(\frac{\mathcal{P}}{\varepsilon})^2-\frac{1}{2\mu}(curl\,\eta)^2\right]$ $\mathcal{H}=\frac{1}{2}\frac{\mathcal{P}^2}{\varepsilon}+\frac{1}{2}\frac{(curl\,\eta)^2}{\mu}$	As a further option, $\eta(x,y,z,t)$ may be considered function of A_1,A_2,A_3,A_4, i.e. $\eta(A_x,A_y,A_z,j\varphi/c)$ which would lead to a different perspective From $\mathbf{E}=-\frac{\partial \eta}{\partial t}-grad\varphi$, with $grad\varphi$ playing no more role in further proceedings $\dot{\eta}=-\mathbf{E}$; From $\mathcal{P}=\frac{\partial \mathcal{L}}{\partial \dot{\eta}}$, $\mathcal{P}=-\varepsilon\mathbf{E}$, $\dot{\eta}=\frac{\mathcal{P}}{\varepsilon}$	

More on variation calculations from [Bronstein, Semendjajew]

Putting it most simplistically, the scope of the work may be summarized in the following table. The mathematics in between the formalism shown and the first quantum step that proved the existence of the photons is beyond the scope of this repetitorium however. That a measurement of electromagnetic energy could only come to an integer multiple of $h\nu$ (Planck const h times freq. i.e. $\sum_k n_k h\nu_k$) is something importan that should be mentioned.

Table[6]

Classical electrodynamics in the language of mechanics	⇔ (compare)	Quantum electrodynamics
$\eta(A_x, A_y, A_z, j\varphi / c) \equiv \mathbf{A}$, $\mathcal{P} = -\varepsilon\mathbf{E}$		Self-adjoint $\mathbf{A}$, $\mathbf{\Pi}$
$H = \int_V \left(\frac{1}{2}\frac{\mathcal{P}^2}{\varepsilon} + \frac{1}{2}\frac{(curl\ \eta)^2}{\mu} \right) dV$		Diagonal $\mathbf{H} = \int_V \left(\frac{1}{2}\frac{\mathbf{\Pi}^2}{\varepsilon} + \frac{1}{2}\frac{(curl\ \mathbf{A})^2}{\mu} \right) dV$
Maxwell-equation equivalents		

Besides, it may be worth noting that $\mathcal{H} = \frac{1}{2}\frac{\mathcal{P}^2}{\varepsilon} + \frac{1}{2}\frac{(curl\ \eta)^2}{\mu}$ on using $\mathbf{B} = \mu\mathbf{H}$ translates back to the more familiar expression for the energy density reproduced below[7].

§§§

[6] Interchangeability relation (free translation from German *Vertauschungsrelation*) is a mathematical requirement.

[7] $w = \frac{1}{2}\varepsilon\mathbf{E}^2 + \frac{1}{2}\mu\mathbf{H}^2$ in correspondence with $\mathbf{E} = -\mu\frac{\partial}{\partial t}\mathbf{A} - grad\varphi$ and $\mathbf{H} = curl\ \mathbf{A}$, Cf. antenna calculations for example. It is however noted from above that $\mathbf{E} = -\frac{\partial}{\partial t}\mathbf{A} - grad\varphi$ and $\mathbf{B} = curl\ \mathbf{A}$. The subtle difference came from the II. Maxwell Eq. $curl\,\mathbf{E} = -\mu\frac{\partial}{\partial t}\mathbf{H}$ or $curl\,\mathbf{E} = -\frac{\partial}{\partial t}\mathbf{B}$ for which one is, respectively, free to write $curl\left(\mathbf{E} + \mu\frac{\partial}{\partial t}\mathbf{A}\right) = 0$ or $curl\left(\mathbf{E} + \frac{\partial}{\partial t}\mathbf{A}\right) = 0$. By the identity $curl\ grad \equiv 0$, both ended up in the respective expressions for $\mathbf{E}$ shown.

Appendices

A Comment on the Transfer Function *H(f,t)*

By T. Vu-Dinh
The author is with 2BT, Melbourne, Australia
Reproduced from TFuncRev03.pdf, 1997, latest revision 2005.

Summary: Fundamentally, the function $H(f,t)$ stated in [2] cannot be justified, with regard to the input/output relationship that it purports to describe.

The transfer function $H(f,t)$ stated in [2], sometimes written as $Y(f,t)$ or more recently in [4] as $Y(f,t,u)$, purports to be the linear time-variant equivalent of its linear time-invariant counterpart to produce the output spectrum by a simple multiplication with the input spectrum. The output spectrum "varies with time" via the time variable[8] in $H(f,t)$, as claimed by A.L. Martin. In the language of linear integral operators, the operation described (a multiplication) is accommodated by the kernel

$$\{M\} = H(f,t)\delta(f-\sigma)$$

Notice the three variables f,σ,t for M. None of the kernels known exhibits more than two variables. This alone would lead to the presumption that M cannot be related to the kernels known. Attempting to formally reduce the number of independent variables to two would lead to nowhere. In particular, it is clear that[9]:

$$F(F^{-1}M) \neq M$$

where $\{F\} = \exp(-j2\pi ft)$ and $\{F^{-1}\} = \exp(+j2\pi ft)$ are, respectively, the kernel associated with the Fourier transform and that associated with its inverse.
Yet, this holds:

$$(FF^{-1})M = M$$

That is, M violates the associative property of linear integral operators and thus cannot be related to the four general kernels known (Cf. Eq.(1) in [1]). In other words, $H(f,t)$ exists nowhere in the function spaces known. More precisely, $H(f,t)$ is not member of a known set, nor member of a set derived from a known set, on which a metric or pseudometric can be defined. In contrast, all functions defined in [3] can be related to the four general kernels mentioned; Cf. [1] for a few typical examples.
The frequency f arises from an integration over the delay in [1] and elsewhere [3], not over the time t. A proper interpretation for $F^{-1}M$ would, therefore, be that the resulting time domain be different from t, ensuring the associative property for $FF^{-1}M$. The fundamental question as to *how* $H(f,t)$ *relates an input to an output* would still remain unanswered however, which is a problem with $H(f,t)$.
It is unlikely that the IEEE-reviewers at the time of approving [2] then were unaware of such fundamental issues, given that [3] had

---

[8] In particular $V(f,t) = H(f,t)U(f)$ where $V(.,.)$ and $U(.)$ are respectively output spectrum and input spectrum. The IBAD quantity, Eq.1, in [2] was derived from this assumption. Obviously, the output spectrum varies with real-time. The output signal $v(t)$ would be obtained (in error) by integration over the frequency, transforming $f \rightarrow t$ where t is the real-time variable.

[9] Consider the inverse Fourier transformation $F^{-1}M$, transforming $f \rightarrow \tau$. To start with, the notation τ has carefully been chosen to be different from t. Following the reasoning of the Footnote above, τ is the real-time variable.

$$\{F^{-1}M\} = \int e^{+j2\pi x\tau} H(x,t)\delta(x-\sigma)dx$$
$$= e^{+j2\pi\sigma\tau} H(\sigma,t)$$

There are two interpretations for the variable t:

1.) t is not the same as τ. This is a proper interpretation for the variable t, which however contradicts the claim that the output spectrum varies with real-time via the time variable in $H(f,t)$, considering that t is not the real-time variable in this interpretation. Notice the three variables σ,τ,t for the resulting kernel.

2.) t is the same as τ, reducing the number of variables to two, the two variables being σ and t. This however leads to the following:

$$\{F(F^{-1}M)\} = \int e^{-j2\pi(f-\sigma)x} H(\sigma,x)dx$$

The latter expression is simply the Fourier transform of $H(\sigma,t)$ with respect to the time variable, evaluated at $f-\sigma$. The time variable vanishes from the result. The result is therefore by no means $H(f,t)\delta(f-\sigma)$. That is $F(F^{-1}M) \neq M$.

been published more than two decades earlier. That, however, the flaw did escape the IEEE-reviewers' attention does not change any of the details shown, nor does that make $H(f,t)$ defensible. The problem was recognized in an early stage of the work of [1] so that in an unsuccessful attempt to map $H(f,t)$ to the function spaces known, the time variable of that function was carefully relegated to the index n as described in [1]. The results of [1] are valid within the limitations imposed by the slowly time-varying channel assumption, ie. nearly constant channel parameters during the $n-th$ observation interval over which the short-time integration takes place, giving rise to the instantaneous spectra $Y_n(f)$. Such an integration[10], however, bears little relation to $H(f,t)$ and its extension $Y(f,t,u)$, both of which are indefensible. The claim by A.L. Martin can thus not be supported. Consequently, the notation $H(f,t)$ was never used in [1], in fact.

References

[1] A.L. Martin and T. Vu-Dinh, "A statistical characterization of point-to-point microwave links using biased Rayleigh distributions", *IEEE Trans. on Ant. and Prop.*, Vol.45, No. 5, May 1997, pages 806ff.

[2] A.L. Martin, "Dispersion signatures, a statistically based, dynamic, digital microwave radio system measurement technique", *IEEE Trans. Select. Areas Commun.*, Vol. SAC-5, April 1987, pages 427ff.

[3] P.A. Bello, "Characterization of randomly time-variant linear channels", *IEEE Trans. on Commun. Systems*, Vol.CS-11, Dec. 1963, pp. 360-393.

[4] A.L. Martin, Tele-IP Draft input to ITU-R F.1093, V1.3, Oct 2001.

[10] Referring to [4], the inclusion of an extra time variable, u, in $Y(f,t,u)$ has added even more confusion to the matter. Besides, radio equipment designs based on the short-time integration concept are the subject for much debate, given the wealth of alternatives available from [3].

A casual look at science and engineering

(Intended to provoke a smile, not to be taken too scientifically)
Reproduced from Sci_Eng.pdf, 2011.

"*Philosophy is the study of general and fundamental problems concerning matters such as existence, knowledge, values, reason, mind…*": From Wikipedia, the free encyclopaedia. *Knowledge* and *reason* are fundamental to engineering and science.

> Did you know that electromagnetic force is "…*1.0E+39*[Note 1] *times stronger than gravity*…"? I didn't know that. The figure caught me by surprise. I picked up the figure from Brian Greene's *The Elegant Universe*, a 3-part series on television that explained efforts to unify EM field theory with gravity and atom physics (strong & weak forces), leading to "string theory". I watched the series intensively, going away with more questions than answers. The graviton? Multidimensions?[Note 2] Parallel universes?[Note 3] Such things were new to me, but the same old questions have remained unanswered: *Where did we come from? Are we alone? Where are we going? What's the meaning of life?*
>
> *What had been there before the Big Bang?* Asking such a question leads to nowhere, for there was no space, nor time before the Big Bang. In particular, in the absence of the time dimension the term "before" or "after" would not make sense. What would happen "after" *a* Big Bang? Such a question would not make sense for an "alien" to ask if *a time dimension* is not known to him. (Replace the "alien" with a human asking the same question. The human is guiding his thought forward along the axis of the time dimension that is yet to be created, an impossible scenario. At the instance of asking the question the human himself is yet to evolve into *existence*, adding to the impossibility of the scenario.) The paragraph serves to highlight limitations of this kind of reasoning. So then, *what are the right questions to ask?*
>
> This paragraph is a concise version of Wikipedia lengthy talk page on the subject. Don't know if the Wikipedia talk page is still available. The preceding paragraph makes this one obsolete. Still, it's interesting to repeat from Wikipedia. What had been there before it all began? Ignoring the previous paragraph, just ask the question anyway. Interestingly, "…*The universe cannot have NO beginning*…", obviously. "…*Yet, at the same time, it cannot have a beginning*…". Had it had a beginning, what had been there before it began? Asking that question recursively will find no beginning for the universe. So then, how *should* we *view* the universe?

Well, go ask mathematicians, physicists and (radio) astronomers for answers to those questions. Mathematicians, physicists and astronomers are admirable and enviable folk, because they are the ones that can help answer those questions, who else. And what does radio astronomy or mathematics have to do with philosophy? Everything, one is tempted to say, to put it most simplistically! It's all about *knowledge* and *reasoning*…

[Note 1]Actually, I remember the figure *1.0E+40* from high school. I am quoting *1.0E+39* here because I want to make reference to Brian Greene's *The Elegant Universe*, just my way of kick-starting the conversation. Because we are at it, which one of the two figures is more accurate? Well, the figures are accurate within an order of magnitude, good enough for everything, why bother asking!

[Note 2]Multi-dimensions? Recall nuclear fission and fusion from physics lectures at school and 1st year university? Think of "our" sun for hydrogen fusion. Either process is accompanied by a loss in mass, say Δm, corresponding to the amount of energy $E=\Delta mc^2$ released into the dimensions obvious to us, x, y, z and the time dimension t.

Now, there is something interesting about the term "missing energy". I remember the term from high school, from radioactive decaying processes. For β-decay, the β-particles (electrons & positrons) showed a statistical distribution for their kinetic energy, suggesting that the Q-value (the loss in mass) was changing statistically from one decaying reaction to another, making no sense of the Q-value. The Q-value must be the same for all decays, because all decaying parent nucleuses had the same initial mass and the reactions must end up with the same daughter nucleuses. The *neutrino* was then invented as an extra product of the decaying reaction to account for the variability of the particles' kinetic energy. It was also interpreted as the means to absorb away the missing energy. "Absorbing away" into where, the physics teacher didn't tell. We school pupils at the time then didn't know to ask the question in the first place. The verification for the existence of the *neutrino* came only after theoretical speculation. I didn't keep myself up-to-date with physics since leaving uni but here is how the decay was written $X \rightarrow Y + \beta + \nu$, with β and ν in their generic form representing both the particles & their respective antiparticle counterparts. The atomic number for the nucleuses may increase or decrease by one depending on the particular reaction. Also, $Q=(m_X - m_Y - m_\beta - m_\nu)c^2$.

If you smash subatomic particles against each other, at speeds near the speed of light, the Q-value for energy conservation does not add up either and some energy "appears to disappear" (Brian Cox's *What on Earth is wrong with Gravity?*). The *graviton* that is supposed to carry away the missing energy is still elusive; The *graviton* is also associated with efforts to map Einstein's spacetime fabric into subatomic dimensions. Physicists in trying to explain the phenomenon have invented new dimensions, into which they say the "missing energy" is released. On top of the four dimensions taught at high school, there are now three extra dimensions (or more) to consider, literally completely out of this world!

Note [3]Parallel universes? Standard material for teaching? Not sure. I stumbled across one journal article on the subject, around 1997/1998. I thought the article was more journalistic than anything else.

§§§

Index

References

field_comp.pdf, 2017.

K. Simonyi, *"Theoretische Elektrotechnik"*, Verlag der Wissenschaften, Berlin 1973.

Küpfmüller, K., "*Einführung in die Theoretische Elektrotechnik*", Berlin Göttingen Heidelberg, 1966.

Wolfgang Gröbner & Nikolaus Hofreiter, *"Integraltafel, Zweiter Teil, Bestimmte Integrale"*, Springer Verlag, Wien Österreich, 1950, 1958, 1961, 1966, revision of 1973.

Bronstein, I., Semendjajew, K., "*Taschenbuch der Mathematik*", Harry Deutsch Verlag, Zürich und Frankfurt/Main, 1974, 14th edition. German translation from Russian by Dr. V. Ziegler, Leipzig.

E.C. Jordan,Editor in Chief, "*Reference Data for Engineers*", 7th Ed., Howard Sams & Co., USA, 1985.

TFuncRev03.pdf, 1997, latest revision 2005.

[References as listed in the comment]

Sci_Eng.pdf, 2011.

[Wikipedia webpage, a.o. as mentioned in the text]